T0293469

Immune System Modelling and Simulation

Immune System Modelling and Simulation

Filippo Castiglione
Institute for Applied Mathematics
National Research Council of Italy
Rome, Italy

and

Franco Celada
Department of Medicine, Division of Rheumatology
New York University School of Medicine and
NYU Hospital for Joint Diseases
New York, New York
USA

CRC Press is an imprint of the
Taylor & Francis Group, an **informa** business
A SCIENCE PUBLISHERS BOOK

CRC Press
Taylor & Francis Group
6000 Broken Sound Parkway NW, Suite 300
Boca Raton, FL 33487-2742

Printed on acid-free paper
Version Date: 20141125

International Standard Book Number-13: 978-1-4665-9748-8 (Hardback)

Library of Congress Cataloging-in-Publication Data

Castiglione, Filippo, 1969- , author.
Immune system modelling and simulation / Filippo Castiglione, Franco Celada.
p. ; cm.
Includes bibliographical references and index.
ISBN 978-1-4665-9748-8 (hardcover : alk. paper)
I. Celada, Franco, author. II. Title.
[DNLM: 1. Models, Immunological. 2. Research. 3. Computer Simulation. 4. Interdisciplinary Communication. QW 520]

QR182.2.I46
616.07'90285--dc23 2014044852

Visit the Taylor & Francis Web site at
http://www.taylorandfrancis.com

and the CRC Press Web site at
http://www.crcpress.com

Preface

This book is about immunology, computers and people. It is an exercise in interdisciplinarity written by an immunologist and a computer scientist who have had the experience of personal interaction that develops between the representatives of different disciplines during the construction of a mathematical or computational model (and by inference, during the writing of a book on modelling biological systems). Interdisciplinarity is an inescapable condition, necessary to keep mathematics securely bolted to the facts of biology and the modellers sufficiently enthused by breakthroughs in immunology to succeed in duplicating these in the circuit of their computer, dynamically correct and easier to test. All this in the hope of dispelling the fog of complexity. As a bonus, the interdisciplinary interaction is known to produce a level of mutual trust and equality that can only be attained in a real friendship.

"What should this lymphocyte do when this antigen appears in the checkerboard?"

"It depends."

"On what?"

"On whether its paratope has an affinity for the epitope or not. Both have eight bits, but only a minority of the B-cells must be allowed to bind."

"I can ask the computer to allow interactions only among those paratopes that have an 8-bit match (0-1 or 1-0)."

"That seems too restrictive: we have to also allow a weaker binding, an imperfect fit. It offers, an *'espace de manoeuvre'*: Nature has the cross reaction."

"Ok. We can say that only the paratopes with 8 or 7 out of 8 matches bind. Do you want more?"

"How many are they now?"

"Let's see … Out of $2^8 = 256$, which is the entire repertoire, only 1 is a perfect fit, and 8 are imperfect by one bit."

"Sounds good. But how can the computer implement high affinity to the perfect and low affinity to the imperfect match?"

"That's easy. I will instruct it to allow the binding for the 8/8 fit every time paratope and epitope meet. In contrast, for the 7/8 fit, for only one time out of ten or fifteen encounters. What do you think?"

"Splendid. For the lymph cells, affinity is the probability of binding! This model is ready to do some runs!"

This conversation took place in 1989, in room 1611 of the Hospital for Joint Diseases of New York, at the precise place where some of these lines are being written today. The interlocutors were an immunologist and an astrophysicist. Franco Celada sparked the idea that contemporary knowledge, combined with the philosophical view that the immune defence is a cognitive system, was so rich as to demand analytical reasoning. This meant transformation into algorithms in the hope of dodging the growing complexity. Philip Seiden, who had just published a discrete model of the formation of galaxies, produced a masterful piece of software.

The model of the humoral response was unveiled in two papers in 1992 and stirred curiosity and interest [1]. Celada and Seiden continued to dabble with their computer model in room 1611, but they also travelled to Santa Fe and to Europe (Milan, Genoa, Florence, Marseille) and were invited to exhibit the model in Princeton. Everywhere, young computer nerds and modellers (e.g., in Milan, Daniele Morpurgo and Riccardo Serenthà, and in Princeton Jeffrey Stewart and Steve Kleinstein) clustered around Philip. He soon realised that in order to promote IMMSIM (IMMune system SIMulator), his code written in APL2 needed to be switched to a better known computer language, and encouraged efforts in this direction. Filippo Castiglione rewrote IMMSIM in ANSI C (and called it C-ImmSim) providing the basis for any updates that may follow: Michele Bezzi enriched this version with the code implementing the cytotoxic response; Kleinstein published a demonstration/educational kit in C++, which was used in the courses of Princeton. Celada and Seiden kept up the momentum by doubling the size and capabilities of the model, incorporating both humoral and cellular response, thus enabling the study of viral infection and response. This version, labelled IMMSIM3, was translated into C (ImmsimC) by Roberto Puzone, and then used in Genoa and New York by engaging graduate students (Claudia Calcagno, Dario Ghersi) and PhDs (Olivier Lefevre, Brynja Kohler, Yiming Cheng and Yanthe Pearson). Filippo Castiglione, with Massimo Bernaschi and Sauro Succi, developed ParImm, a version of C-ImmSim able to run on parallel computers. Later, C-ImmSim branched into SimTriplex, a customised version to simulate cancer immunoprevention, which was further developed by Santo Motta and Francesco Pappalardo. The exchange of information among groups has been frenetic, and the need for testing and refining overwhelming.

Philip Seiden stood by all new versions and smiled, happy. But in 2001, he could not attend a month-long workshop on IMMSIM that we planned in Bielefeld, Germany. He was 67 when he passed away. What a loss for all of us.

This book is an attempt to show, from two perspectives, the result of such interdisciplinary effort, and to give a measure of the intrinsic difficulties we have had to face at the beginning problem. The differences in the scientific language used were just the first of this tricky business. The lack of a reference work was another: IMMSIM was the first of its kind.

How does one experiment with interdisciplinarity in the writing of a book? We, the authors, decided to feel free to do whatever we deemed useful, for whoever might be attracted by the idea of dabbling in a discrete modelling of the immune system. Celada chose to tell the story of the adaptive response as if the readers where aliens coming from another galaxy (first part), whereas Castiglione revealed the secret steps and ways to create and use a model like C-ImmSim, something that he knows better than anybody else, as if the readers were humans, and probably immunologists (second part).

This approach may not be perfect but it fits our expertise and our respective 'offspring'. No wonder; IMMSIM/Immsim-C and C-ImmSim are different algorithms, are coded in more than one language, run on different machines, but they implement the *same* body of ideas, choices, concepts and philosophy, that we refer to as the Celada-Seiden model.

Filippo Castiglione, Rome
Franco Celada, New York

January 2015

Acknowledgments

We would like to express our gratitude to Ann Rupel for her professional work of scientific editing and precious assistance in finalising this book.

Secondly we would also like to thank our families to whom we wish to dedicate this book.

Contents

1

Immunology for Aliens

It is intriguing to write for, and about, a population of researchers who are sometimes called '*aliens*' by the classic immunologists. Nobody knows from where they came, but the point in time of their appearance can be traced to around the last decades of the twentieth century. They are still around, and some have proliferated. They are also called *modellers*, but they are fastidious about what they like to model. As it happened, most of these modellers developed their tools and formed their models aiming to understand aspects and parts of the immune response. They were eager to learn all they could about how immunity worked, and their immunologist friends were happy to tell them what they knew and what they believed. The picture that the aliens received was very 'cellular'; individual cells developing, maturing, moving around, meeting and conversing with other kinds of lymphocytes, and finally producing incredible responses. This picture may have been slightly biased by the way the immunologists used to communicate: they thought that the best way to convey their knowledge to the aliens was to do it in the form of a fairy tale, perhaps because they just liked fairy tales or because they were all admirers of the character played by Francois Truffaut in the contemporary film Close Encounters of the Third Kind, the man who tried to facilitate communication between inhabitants of different galaxies by using music.

Philip Seiden was the first alien who came to the immunology lab (of Franco Celada, in 1990). He introduced himself as an astrophysicist, and said he was working at IBM in a place that sounded like 'Sing Sing'. He listened intensely, pondered, then made up his mind: the best way to represent, and therefore to observe and understand the immune response, was to mimic it as a *discrete model* on the computer. At that time, models of this kind were called 'cellular automata' where the adjective alluded to space divided into tiny chambers, and not to the biological cells that were being represented. Today, the saga continues, but the most accepted definition of what goes on in the computer is an 'agent-based model'.

An immunologist and an alien are writing this book together. The situation is reminiscent of that of 1990. The aim of their effort is not only to show and explain what happened, but also to be a useful example to researchers willing to enter into the world of modelling. There is no reason to abandon the fairy tale tool that has been successful once, based on the aliens' metaphor. In a number of informal paragraphs, most of the facts and concepts deemed central to the immune system are touched upon: its birth, its evolution, its functions and its breakdown. In the following paragraphs, the concepts that have inspired IMMSIM and the results of IMMSIM which have helped clarify and test those ideas and concepts are singled out and, hopefully, made clear.

1.1 Leo Szilard Paradox: Grand Central Terminal

In 1948 the atomic scientist Leo Szilard published 'Report on Grand Central Terminal' [2] about aliens landing on Earth after a nuclear holocaust has wiped out life on the planet. The brilliant but underappreciated young physicist, Xram, traces the cause of that intercontinental war to the finding of metal-disk-operated locks in the toilet doors in the Grand Central Station of Manhattan (Figure 1). In his theory, the eagerness to obtain the disks to be ritually sacrificed in the door locks makes these

Figure 1. Xram inspecting the metal-disk-operated lock of a toilet door in the Grand Central Station.

objects very valuable, as their possession permits one to carry out an urgent act, be it a cult-related action or a more bodily relief. Anyway, it is the first step in a chain of greed-triggered events, ultimately leading to an intra-species nuclear conflict. Xram's theory is met with scepticism by his alien colleagues, who remain in the dark about the cause of wars. However they continue to struggle in their all-round exploration of what may have been life on Earth.

Now seems the right time to comment upon and discuss the principles of adaptive response and immunological memory, where the main trigger is not greed, but survival, at all levels. (In the words of Francois Jacob, "The dream of a bacterium is to become two bacteria," and everybody knows that the dream of a cell is to become a clone.)

Note of caution: Aliens, with their pixel-equipped sensors, do not read concepts in sequence. They take chunks of writing up to twelve lines long, to be ordered and elaborated later. They also like repetitions. They will be satisfied.

1.2 Non-specific Defence

When life on this planet takes the form of multi-cellular associations, the space is automatically separated into *inside* and *outside* of each organism, and in order to survive, each has to come up with ways to keep the inside from being invaded by outside dwellers. Besides physical barriers (skin, epithelia, membranes), groups of cells become dedicated to the riddance of trespassers, bacteria, fungi, and protozoa. The defender cells will recognise them as not part of the organism by chemical differences pertaining, for example, to the membrane structure. They internalise them by phagocytosis,[1] and kill and digest them with enzymes. However, they are not able to distinguish one invader from another, therefore they have no *specific* memory of a given invasion that would help them to be prepared if the same invader should come back another time. Responses remain exclusively *non-specific* for two billion years. Only then, new capabilities allow the introduction of specific, also called adaptive, responses. They are added, together with their specific memory, and in part, spliced to the previous ones, instead of replacing them.

1.3 Levels of Evolution ♞[2]

Like individuals with respect to species, cells in multicellular organisms are disposable commodities. But the fate of the single cells depends on the role they may play in the life of the organism: in case a trait, produced, for example, by genetic mutation, makes one cell's role critical or life-saving for the organism, the new trait will be passed to the organism's offspring, the role of the particular cell will be reinforced, and gradually with time

[1] Words requiring explanation can be found in 'An unconventional glossary for aliens' (page 70), where they are catalogued according to the section in which they appear.

[2] The meaning of the relation codes on titles is explained in Table 2.

and survivor's luck, the new trait becomes characteristic of a majority of the population.

1.4 How Is a Trait Selected If It was Not Needed?

It is more difficult to imagine how individual changes become assets of the species when the effect of the new trait is not dramatic and not apparently life-saving. Or not even noticed for a period of time. This is the case of adaptive immunity 500 million years ago. Multicellular life in the primitive ocean was thriving and complex; sometimes enormous, molluscs and crustaceans did not seem to need a drastic change in their defences, which were non-specific. One definite reason was the unchanging, low temperature of the water, which limited the rate of multiplication of micro-organisms (speed of growth doubles every 7°C rise in temperature). When, in the early vertebrates, lymphocytes appear and begin to produce immunoglobulin-class structures, the moment of *need* for specific responses is still three hundred thousand years away in the future.

1.5 Luck or Foresight?

Possibly, the new lymphocyte came as a package with another modification. The other item was selected because it was of immediate use, and the lymphocyte managed to piggyback; perhaps it was selected by mistake or at least without an apparent reason. The later vertebrates should be thankful for that, as it allowed them to transit eventually onto the dry land and use warm blood, whereas most arthropods and molluscs failed. Incidentally, a few days after the preceding sentence was written, the media showed an example of 'gratitude' inflicted on a specimen of sea lamprey that was captured near the coast of New Jersey (Figure 2, left panel). Some local newspaper called it a 'fearful mutant' perhaps considering that it was living in one of the most polluted oceans of the world.

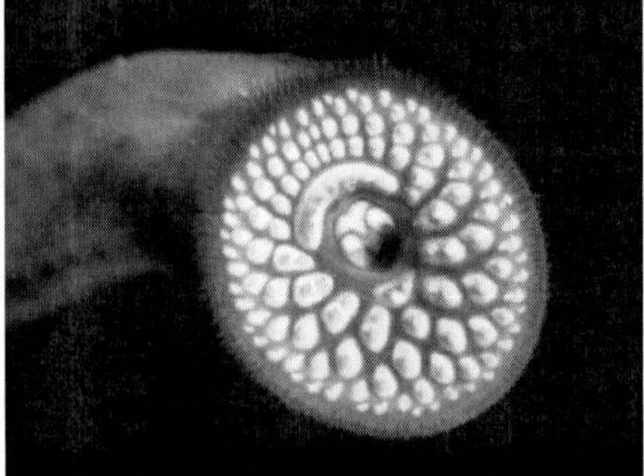

Figure 2. The lamprey captured near the coast of New Jersey (left) and its unusual mouth (right). Reproduction with permission from the author Doug Cutler.

Colour image of this figure appears in the Colour plate section at the end of the book.

1.6 Blood Brothers ♞

A solution to the puzzle (How could the adaptive immune system enjoy an evolutionary advantage long before the conquest of dry land made it vital for the vertebrates?) comes to mind after considering the lamprey's lifestyle and the structure of its mouth (right panel of Figure 2). This animal has a most efficient way to vacuum-connect to the skin of large and more evolved fishes, break into their veins, suck their blood and spend most of its life as their unwanted appendices. Both partners have adaptive defences, which include graft rejection capability, which constitutes a barrier against colonization by foreign cells, and chimerism. This means that thanks to immune protection, the minor partner will be able to go on sucking blood for years without the fear that its genetic identity may be endangered by symbiosis with the host cells. A tentative hypothesis is that this protection of its identity has helped the lamprey in a competitive environment and paid the ticket for the transit of the adaptive immune system to life outside the ocean (where, at last, it became pivotal for the defence of the new vertebrates). In the course of evolution, lampreys and other pilot fish followed a long line of older blood-sucking beings: among others, molluscs, arthropods and a myriad of insects. How could these species thrive without immunocytes? Either the evolutionary distance between sucker and sucked was too large to permit intimate cell mixtures, or the time of contact was too short to jeopardise their identity (Figure 3). With the advent of man, bloodsuckers in general have a hard time; they are feared, slapped, even sprayed with some of the worst poisons. *Anopheles* gets the well-deserved label of 'Enemy number one', for its role in spreading malaria. Leeches have mixed fortunes: for two thousand years they were the most valued weapon of medicine but they are now either forgotten or culturally linked to a certain count in Transylvania.

Once again, it is the lamprey that must be mentioned. Immunologists will consider the symbiosis enacted between

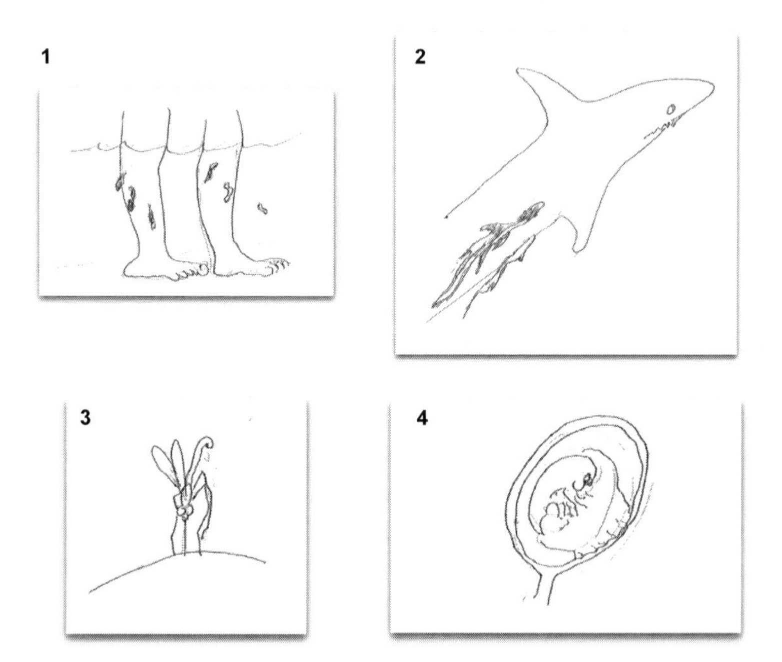

Figure 3. Here is a collection of four bloodsuckers: 1) the leech, 2) the lamprey, 3) the mosquito (*Anopheles*), and 4) the foetus in the womb. The four 'sucked' are in the order: man, shark, man again, and characteristically, woman. Looking at these examples, note that the partnership is long lasting in vignettes 2 and 4, while it is occasional in vignettes 1 and 3. This is to suggest that possessing an adaptive immune system is a protection against loss of genetic identity during long term 'cohabitation' of heterologous cells.

fish of different species as a dress rehearsal for the evolution of mammals. Why? The vital contact between uterus and placenta takes place between two organisms with specific immune systems. Each of them needs to protect its identity against the invasion by the other's cells and for this, the presence of their two adaptive lymphoid systems is a godsend. On the other hand, the rejection of the foetus by the mother would be such a disaster for the species, as to justify drastic measures (such as hiding and even changing the nature of surface determinants of the trophoblasts to avoid it). Finally *specific tolerance*, which is a function of the immune system, is attained, even if it looks more like a sequel of compromises. The larger of the two organisms tolerates the presence of the foetus, the real bloodsucker, for nine months.

1.7 The Members of the Winning Team are Rewarded ♞♜♝

Here is how the specific lymphocyte is *actually* rewarded: binding the antigen prolongs the cell's life. Even better, the trademark act of cell-cell cooperation, e.g., between helper and effector cells (T-cells and B-cells), or between activated dendritic and effector cells, is decisive in allowing the cell to become a clone. To give an example, when the B-cell has completed the binding, internalisation and processing of the antigen, and the helper T-cell (thanks to the ways and means provided by the lymphoid organ facilities) has made contact via its T-cell receptor (TCR) with the peptide presented by B-cells' MHCs (major histocompatibility complex molecules), both cells put out lymphokine receptors. Then, the helper saturates the near environment with short range messenger molecules; these bind the receptors, and transmit the impulse; both cells become *activated* and set up procedures to produce clones.

1.8 Somatic Recombination Produces Diversity among Equals ♜

The adaptive response relies on *specificity*, and specificity relies on distinctions within cell populations that share functions. These populations are coded by germ line genes somatically recombined in each maturing B or cytotoxic T lymphocyte (Tc). As a result, the antibodies (Ab) are all identical, except for the paratope, which is the part that makes contact with the antigen (specifically, with a reciprocal part of the antigen known as the *epitope*). The *para-epi* binding is the first and most necessary event which initiates the organism's defence against any invader. What ensures that even a totally unknown invader will be tagged, is the large *diversity* and number of cells available to generate the large repertoire, that is, the almost certain availability of the specificity needed for any defence. To assemble the paratope, the cells utilise certain preformed genetic sequences that are chosen randomly from large collections, and then spliced together. Both the selection of the pieces, and the splicing by enzymes in specialised organelles, are processes where stochasticity is prominent, and therefore it is impossible that two maturing, similar, apparently identical cells turn out identical paratopes. More thoughts about this in Section 1.10.

1.9 The Workshop of the Adaptive Response ♞ ♜

It is mind-boggling to consider how many new mechanisms and structures would be needed, ready and all at once, to start off the new *adaptive defence* under stress. It is thanks to the efficiency and resilience of the old system, that the new system was allowed to develop at its own pace, protected from an incumbent necessity. This is another way of saying that the adaptive response emerged long before it was really needed. The non-specific immunity not only kept the lampreys out of danger while they were experimenting with the revolutionary new deal, but also furnished key functions to the new responses. The non-specific phagocytes in particular, well known and much appreciated for contacting, encompassing and internalising any external particle or object they could reach, kept adding incredibly sophisticated capacities to their original work specifications. The object that at first was simply captured and kept away, was, with the passing of ages, destroyed by enzymatic digestion. An additional step in evolution made these cells able to enter into collaborative activities with the lymphocytes, when they began to sort—by size and shape—molecules resulting from digestion and expose them to newly formed structures of the cell membrane, specifically contacting certain receptors of the collaborating cells. At this point of evolution, they became a key figure for the new system, the 'antigen-presenting cell' that did much more than present antigens, as will become clear when discussing activation (see Section 1.22).

1.10 Repertoires ♞♜

Diversity, as stated earlier, is the speciality of the lymphocytes. Each of them synthesises its personal paratopes by recombining preformed segments of germ line DNA: the number of segments used (two or three), the alternatives in size and the random variations of the segment joining generate a very large number of possibilities (the number of different paratopes using 20 amino acids could, if all 20 existing options appeared only once, reach the order of 10^{18}, which is enormous and difficult to grasp, even if a fraction of the diversity may never materialise because of identity of functions in different amino acids). Consensus at a lower order (10^{11}) can be reached by estimating the variability of the three hyper-variable regions (HVR 1, 2, 3) in the variable regions of the antibodies, and adding the variability produced by random errors in the splicing together of the coding segments. The discrepancy between estimates of diversity is upsetting, but cannot be remedied by deeper knowledge of statistics, since it is caused by elements of biological complexity. Lymphocytes constitute a large source of the diversity of any organism, and it is easy to accept that no two lymphocytes, even under identical circumstances, will ever produce the same antibody or the same specific receptor. The evolutionary significance of this is that the assembly of germ line information in the paratopes is a somatic deed and this introduces stochasticity. A significant feature of the system is that it lets specificity be managed by single cells which, as a rule, carry just one type of receptor and therefore one specificity. This is a great solution, but is also a limitation in the expression of full immune potential. In man, lymphocytes are probably in the order of a billion, and account for half of the entire weight of the lymphoid tissue, but many of them are part of clones (e.g., in the case of memory).

1.11 A Battalion of One ♞ ♝

Even if there are some corrective factors, such as pre-stimulation cell division, and cross-reaction (which allows low-affinity cells to respond), the *limited* number, per clone, of naïve lymphocytes is a constant feature of specific responses. It is exacerbated by the large availability of epitopes to be covered at any intervention, and the rule that limits to one, the specificities of one lymphocyte. The first consequence of this situation is that no immediate immune response is possible in the naïve animal encountering an invader for the first time; instead, a sufficient cell population must be built up by proliferation, and this takes time, even if the cells are properly stimulated and the cells are duplicating every six hours. The time taken to reach critical level contrasts with the rapid deployment of either the innate or the memory responses and is at a disadvantage with the speedy growth of the external invasion.

1.12 Self-inflicted Damage Threatens Survival ♜

The new specific immune response mounted by the lampreys is endowed with the capacity of speedy expansion and the use of potent effectors; this makes it a weapon without a safety catch. The policy of adaptive defence is to encourage the formation

Table 1. Different phenotypic traits and their inheritance. The question mark about inheritance of items in line 3 (paratopes and receptors) must be qualified. All three CDR (Complementarity determining regions) of a paratope contain coding material *inherited* from the *germ line*, and therefore can be considered a legacy of evolution, making sure that individuals of today are equipped with specificities that have belonged to the fittest vertebrates of the paleohistory. The elaborate mechanisms of this transmission seem to favour this interpretation. On the other hand, there is reason for scepticism: the complex rearrangement by soma may have altered the original specificity beyond recognition, and the epitope panorama of the Mesozoic era may have changed beyond recognition.

1	Fingerprints	Pure Soma	NOT INHERITED
2	ABO and HLA	Pure Germ line	INHERITED
3	Paratope and TCR	Germ line rearranged in soma	? : A PUZZLE

of the largest possible repertoires; the method adopted is the generation of molecules by recombination, in the individual lymphocyte, of a number of codons derived from the genetic germ line (and therefore from the evolutionary inheritance), a method that results in a large diversity by a variety of mixing and gluing imprecisions. There is both a strength and a weakness in this method. The strength is the large number of individual specific cells it produces, ready to be tested for their capacity to bind the epitope of the attacking or potentially dangerous antigen. The weakness is the minuteness of the fraction of the attacker that the responder contacts and perceives: 3 or 4 amino acids at most, or perhaps six sugar units, a really tiny portion, one that reminds one of a key with few cuts, that nobody would be surprised if it opened the wrong door. The wrong door in this particular metaphor is any *self-molecule* in the body, and an attack to self may be, or may rapidly become, lethal. It is futile to criticise nature, by showing through numbers, how more secure would be a key with ten cuts, or an antibody with a paratope capable to bind stretches of antigen ten amino acids long. Suffice it, for the record, to note that 'only' 8,000 different epitopes can be constructed by combining 3 amino acids, and almost 160,000 by using 4. Obviously, small numbers here indicate a high probability of overlap and cross-reaction between self-antigens and anti non-self responses.

1.13 Don't Think, But Reason and Understand

As expected, the newly born adaptive system tried neither to increase size and/or precision of paratopes, nor to encompass larger epitopes to increase security. Instead, it surprised a generation of immunologists with a large stride toward cognition: the danger of mistaking self for invader will decrease only if the system improves its knowledge of any antigen it encounters. Things around this time must have been hectic, because the early vertebrates were in real danger.

In the *next era*, the antigen continues to be contacted by receptors binding epitopes of 3–4 amino acids, but now, before launching a response, the rest of the antigen molecule (with a molecular weight of at least 10,000 amu [atomic mass unit] and without upper limits) has also to be examined. For this task the macrophage "old type" cells, able to internalise antigen upon contact, are engaged. This task is carried out, for antibody responses, by the B lymphocyte, but for T responses by an antigen-presenting cell of the macrophage family, the dendritic cell (Dc). After internalisation, the antigen molecule is subjected to a thorough enzyme treatment that dismembers it; a few samples of the molecule, in the form of 14-amino-acid-long peptides, are mounted on MHC class-2 receptors on the cell's surface, whose aim is to communicate with the helper T-cells (Th).

This is the right point to state that the key of *new era* arrangements is the prohibition of initiating any humoral or cellular immune response if it is not specifically approved. It is the helper T-cells that will examine the nature of the proposed antigen, and make sure that no self-aggression takes place. When a Th cell connects with the loaded class-2 MHC, it must decide whether to issue a permit in the form of a positive stimulation via cytokines or to deny it. There is a catch: a fraction of the potential Th examiners may not be available. As explained in section 1.17, those T-cells which are specific and potentially aggressive for self will not be available because they have been eliminated in the thymus. Therefore the terms of the decision must be rephrased: either, "Yes, go!" or a deadly silence, duly interpreted by the effector cell as a, "No". The effector in this case, be it a B-cell or cytotoxic T-cell, does nothing, and dies.

The logical construction of this decision is splendidly lucid/ transparent/explicit: to assure accuracy, it disregards possible overlaps at the level of the tiny epitopes, but relies on the pre-selected, pre-certified corps of judges. Brilliant! (More on cognition in the immune system is found in section 1.35).

1.14 Full Cover Comes at a Cost ♞ ♝

This success is reached at the cost of precision, and the picture of the antigen is rather blurred. There is no doubt that the *strength* of the interaction has merit: ideal binding occurs when the shapes of paratope and epitope match perfectly, allowing the active groups (electrically charged or hydrophobic) in epitopes and paratopes to be positioned at a short distance from each other. However, and this is a fundamental compromise, bindings of lesser perfection and thus lesser stability are also allowed, making it possible for a single paratope to recognise an entire population of antigens, and a single antigen to be recognisable by a population of diverse lymphocytes. This trick is called cross-reaction, and extends the reach of cells by several orders of magnitude. On the dark side, a cell specific for a foreign antigen can bind to the *same* group of three amino acids residing on a *self-molecule*, and if no default mechanism is enacted, this mistake can have serious consequences.

1.15 A Bet, Not a Fantasy, about Earliest Happenings

According to the colourful plot in Figure 4, the specific response started with a relatively low level of complexity but a high level of efficiency. Even if one cannot exclude the early presence of threats by viruses, their low complexity is attributed to an initially just-humoral response, with one of its fields of action being bacteria. Success in this case would have accumulated

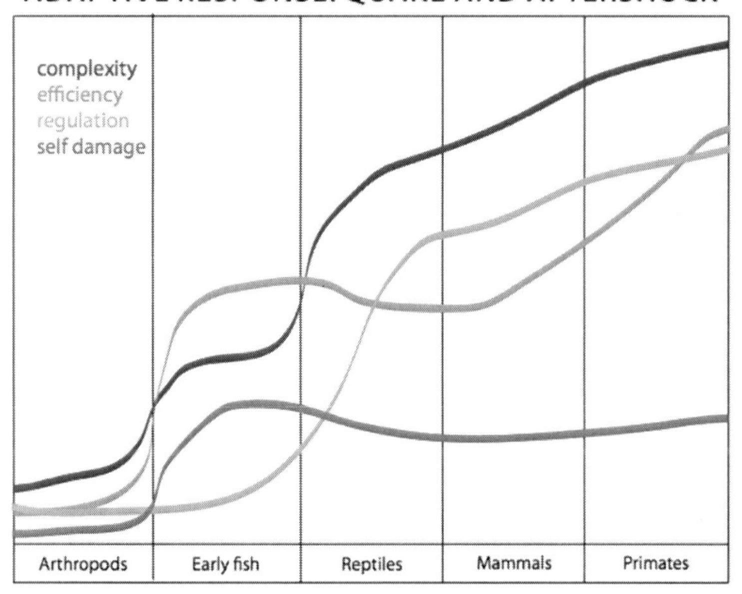

Figure 4. (Artist: Valentina Celada) A 'paleohistory' of the immune system: arthropods and molluscs of today exhibit the oldest kind of defence, non-specific, cellular and humoral that dates back 800 to 900 million years, and has allowed multicellular life to populate the primitive *bouillon de culture*. About 500 to 600 million years ago, *early fish* form tiny vertebrae and are surprised by one of the major revolutions in biology; thanks to lymphoid cells, the response becomes specific and the diversity grows at all levels and in repertoires. Responses are effective, but friendly fire (auto-aggression) is staggering. Reptiles, 300 million years ago, and then the dinosaurs, inherit a faulty weapon that shoots also backward. An immunological cause for the extinction of the latter is difficult to prove, but even more to disprove. Mammals, at 100 million years ago, build up the system in the only evolutionarily correct way: every cell, every clone stimulated is controlled and regulated (killed) if it shows auto-aggression. The immune system becomes multi-organ, allowing for special rooms for cells to meet and kill the suspected. And primates? Well, primates, appearing about 60 million years ago, must have been engaged in the last step of immune evolution, which is fully active even today. Memory cells have become central in the defence, because they are enjoying large numbers and ready helper cooperation. However memory thwarts new specific responses and this is against the trend that began in the lamprey. The cross-reactive, low-affinity dominance is arguably the cause of senescent immunity, whose principal target is man (at 25 million years ago), and especially modern man whose life has dramatically increased 100 years ago (see section 1.50).

Colour image of this figure appears in the colour plate section at the end of the book.

evolutionary merits and earned for the new system, the right to continue its experiment, making changes and trying them for fitness. The higher chance of survival of the organism carrying the trait becomes the blue stamp for further evolution. The damages produced, by-products of a 'purely effector' or 'trigger ready' immune system, initiate the process of finding remedies and this entails a large increase of complexity. The size of the remedy and the rise in complexity (in terms of controls implemented) testify to the seriousness of the dangers of this gun without a safety catch. It is easy to identify which dangers were encountered. They were of two natures: autoimmune (specific response, anti-self) and indirect damage (since there is always an indirect danger in a battlefield). This means that too many responses were launched against objectives which were not a threat to the body. In the next paragraph some of the remedies implemented will be described and, no doubt, the description will give the impression of a logical and planned intervention. This impression is wrong. The process must have followed the usual path of evolution *à la Darwin*: large varieties of random mutations, and automatic selection of those associated with survival. Even if nature is known to be a tinkerer and a specialist of re-use, she is neither a miser nor is she thrifty. And for her, time is no object, nor is it a complication.

1.16 Cooperation among Very Different Cells Multiplies the Number of Jobs that Can Be Done ♞

Cell cooperation is any productive interaction between different cells of the immune system. It is the trademark of immune defences and is used systematically. The control over self-aggression by effector lymphocytes was the result of the disappearance of the primitive trigger-happy organism, substituted by those whose effectors were simply unable or unwilling to mount a response in the absence of a 'help' stimulus. One can imagine that the encounter between identical lymphocytes would have been rather boring, while that of a lymphocyte and, say, a macrophage (MA), would have generated many possible types of energy release. Since, by definition, energy can cause various sorts of stimulation resulting in the multiplication of the lymphocyte, it is immediately clear that this encounter could, and did, become extremely useful in mounting the control of the system. The necessity for the two cells to have physical contact to interact properly adds precision to their action, and adds the need, and therefore the opportunity, to guide and manage it by the development of 'the lymphoid organs'. These organs are the lymph nodes, the spleen, the bone marrow, and the remarkable *thymus*. Another property of close encounters is that, as a rule, when one of the two cells meeting is or becomes excited (activated), the state is shared with its counterpart. If the two are lymphocytes, one B, the other Th, or one Tc and the other Th, the shared excitement causes the formation of two clones, one from each. If one of the interlocked cells is an antigen-presenting cell (APC) and the other a lymphocyte, only the latter will clone while the macrophage will keep the excitement and pass it further in the next encounter (see Figure 5).

In the humoral branch, protein molecules are antigens, and they are also potential immunogens (that is, able to elicit a response) provided they are large enough to contain *peptides* (usually hidden from the B/Ab receptors). The size divide is thought to be around 10,000 Daltons. The immunogen's epitope will be bound by a specific BCR, and then endocytosed and enzymatically processed, which makes the peptides available as probes for identifying the status (self vs. nonself) of the prospective immunogen. For the present test, only anti-foreign helper T cells are available, and only *nonself* peptides will be bound and cause the cytokine storm that will trigger the response. This admirable design, however, suffers some inevitable exception. What is described above (see also Figure 5) is called *intramolecular help,* since the element that binds the BCR and the element that is presented to the TCR belong to the same protein molecule, the same covalent chain of amino acids. But there are cases where the epitope and peptide are separated by a *non-covalent bond,* and still the help is delivered. This variety

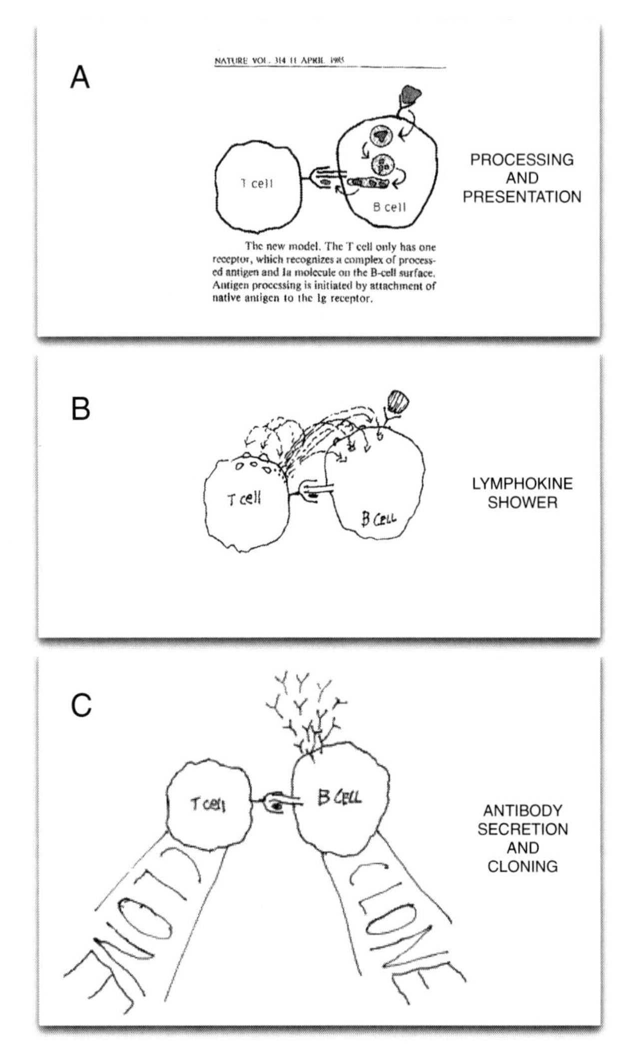

Figure 5. In April 1985, Nature published a letter by Antonio Lanzavecchia, titled 'Antigen specific interaction between T and B-cells' [5]. It described the steps of cellular cooperation, one of the most debated topics of the previous decade. In the 'News and Views' section of the same issue, Jonathan Howard presented the work as 'Immunological help at last!' [6] and provided the figure in panel A, which became an instant classic. The illustration focuses on B-cell binding and processing the antigen, and presenting it to the T-cell. For this book, panel B has been added, showing the initial effect of the T-B interaction, the activation created by means of lymphokines secreted by the T and collected by timely produced receptors on the surface of both cells. Panel C has also been added to show the effect of the lymphokine shower: the two cells get ready to multiply and become clones and the B-cell, ending its hesitation thanks to the T-cell message, secretes the antibodies that she had been synthesising for some time. Bravo. Mission accomplished.

of help is called *inter-molecular help (IMH)* and, for example, takes place when the antigen is a polymer of not-identical units. But where IMH becomes fully significant is in producing help to antigens in *Immune Complexes*. This happens in the case of *Rheumatoid Factor (RF),* which consists of antibodies elicited against some epitope of the FC region of the antibody moiety, being helped by any nonself peptides belonging to *bona fide* antigens against which the immune system is responding. It is an interesting example of *inevitable* anti-self response. RF is prominent in all rheumatic diseases and in lupus, but, as could have been predicted, is traced in all normal responses some time after immune complexes appear. The good news is that in the healthy individuals their RF titer remains low, and no adverse symptoms appear, telling us that to build the autoimmune disease it is not sufficient to have the occasion of intermolecular help. We don't know why, but we are happy nevertheless. Even in Autoimmune diseases RF is more famous as a diagnostic index than as a damaging weapon. Intermolecular help, used as a clever tool, has allowed production of anti-CD4 secondary antibodies after a pulsing with CD4/gp120, implying that the high titers of anti-CD4 in AIDS patients are directed against the receptor-gp120 complex [3]. The IMMSIM team of New York first simulated intermolecular help in 1996 [4].

1.17 Thymus Deeds ♞ ♜

The remarkable *thymus* is the station through which all T lymphocytes, newly born in the *bone marrow*, must pass to mature. It has been called a university built to educate a population of T-cells, but the thymus does not teach, nor does it try to educate or instruct its pupils. It is a machine made to test them and select them according to a number of criteria. It has ways to analyse each candidate for its capacity to relate in terms of specificity and affinity and has the means to determine whether the results indicate that the tested cell is definitely dangerous, absolutely useless or probably useful to the organism. Only those in the last category will be promoted, and this means that 95% of the candidates will never leave the premises: they will be eliminated by apoptosis or die due to neglect. The need for such a drastic selection process is a lesson in Darwinism: in the bone marrow, where it was generated, each T-cell was let free to program the structure of its determinants by combining fragments of preformed genetic information. The procedure is stochastic, not deterministic; it allows a large diversity within the cell population, but has a low yield of acceptable cells, less than 5%. The thymus is a ruthless selector which resembles a university only for the latter's obsession with tests; however it is difficult to do justice to its complexity and richness. The 'graduate cells' comprise T-cells with different functions such as Th, Tc and the more mysterious T regulatory (Treg) cell. Decisions about who dies and who lives on are at the centre of a cognitive mission of the immune system. After the elimination of the self specific ones, the T-cells that graduate out of the thymus endows the body with an invaluable treasure, the processed information about the outside world and its menaces, perceivable as the reflection of the deepest inside, with the police files of all the most dangerous aggressors. To be positively selected, the T-cell's receptor must be able to interact (i.e., bind with low affinity) with the antigen-presenting receptor on thymus macrophages in the absence of an antigen. This qualified self recognition defines the biological individual; the reason for this requirement is that in the future clonal activation of the T-cell, it will have to bind to a self APC presenting in its context, one peptide derived from an invader. Only if the loaded receptor binds with a higher strength than that of the empty one will the response take place. The beauty of this arrangement is that the foreign specificity is perceived as a variation in a benign self/anti-self interaction.

1.18 Hypothesis Is a Bet: Tips for the Aliens ♞ ♝

Everybody likes to watch good body theatre, such as the sensual dance of Pina Bausch's Tanztheater. Now, a microscope is not usually connected with one's computer. Aliens are therefore advised to ask their immunologist friend (say his name is

Antonio), to take them to the bench and perform a T-cell activation experiment in front of their eyes. He will take a culture of lymphocytes and place the *petri dish* over the lens of an inverted Zeiss, and ask you to focus at magnification 40. The cells are swimming graciously, dancing in Brownian motions; they do not touch. Now Antonio adds normal dendritic cells that have been collected from the same donor. The lymphocytes show some interest in the newcomers; when passing near each other, two cells make contact, their membranes seem to move and unfold as they seem to circle each other. They have explored each other thoroughly, but they have not found reason for further interest, and they leave in different directions. Next, Antonio adds another suspension of dendritic cells: they have been exposed for several hours to T-cells purified from the same donor, and this time an antigen, the Tetanus Toxoid (TT), against which the donor was vaccinated when he was a child, is added to the mixture. Even a first timer at the microscope would notice a difference. The dancing and the mutual stroking of the two types of cells picks up with a new energy, and involve so many T-cells that he will suspect that the phenomenon is not entirely specific, or rather, that many of the T-cells involved must be unable to bind seriously to the presented TT. He is right. Only very few specific T-cells receive an input from the nearest presenting dendritic cell when, helped by chemotaxis, they meet. The T-cell, in response, not only secretes lymphokines that create the excited atmosphere of *village sagra*, with everybody dancing around, but also contribute to make its own 'marriage' become a cluster of many apparently active T-cells, even if only one or two of them are specific. The *cluster* image is typical and easy to spot. Counting the clusters in the dish helps evaluate the success of the experiment on the bench. But a vivid memory remains of a meeting in a place called ZiF (Zentrum für Interdisziplinäre Forschung), where Avrion Mitchison of University College London, a legend in the realm of cellular immunology, presented a paper affirming that clustering of eight and more cells is the *sine qua non* condition of the antigenic stimulation of the T-cell.

1.19 To Grow and to Change ♛ ♚

Like life on earth, the immune system relies on growth, and the advent of the lymphoid cells has represented acceleration and sophistication. Under stimulation, lymphocytes can duplicate up to once every six hours, which means a clonal growth of sixteen times in a day, which is extraordinary for mammalian cells (bacteria double by the hour). The rapid increase of specific cells is essential for the efficiency of responses: this is most clearly seen in the defence against infection, and in the 'number effect' of single specific clones in action. Strictly coupled to the mechanism of cell division is the provision for a change, epitomised by somatic mutation. Change itself is ambiguous;

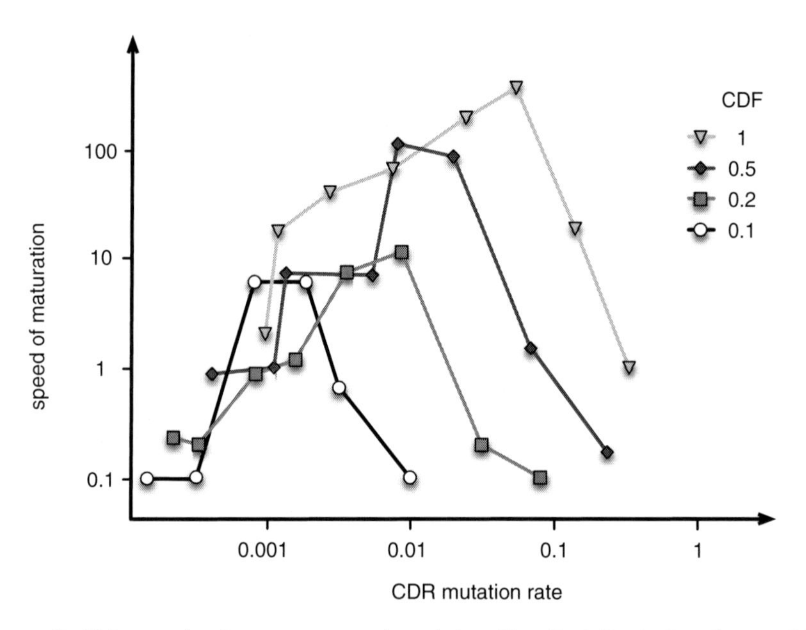

Figure 6. This graph shows an experiment *in silico* that illustrates the positive-negative effect on affinity maturation observed when the mutation rate increases. The experiment is run in IMMSIM; in the abscissa is the complementarity-determining region (CDR) mutation rate, growing over three orders of magnitude (from 10^{-3} to 1 event per time unit), in the ordinate is the speed of maturation of the affinity. The plot combines four similar experiments differing for the fraction of mutations in the chromosome that actually affects the CDR (This is called the CDR fraction or CDF) (abridged from [7]).

Colour image of this figure appears in the colour plate section at the end of the book.

it can be either an occasion for the evolution, expansion and success of the clone, or the cause of its loss of character, function and efficiency. Obviously, either effect depends on mutation rates, and on the efficiency of repair mechanisms; the range of physiological rates is selected by evolution on the basis of survival, or specific success. In the clonal progeny of a B-cell, the focus is on changes in binding capacity induced by mutations in the antibody coding genes. Any single event may cause an increase or decrease of the affinity of a cell in the clone (or even annihilate it altogether), and any change, regardless of the immediate utilisation, materialises as an increase of diversity of the clone and of the organism and serves as a handle for further progress.

Mutation triggered in the context of the response is a very interesting phenomenon and should be kept separate from random mutation, both in terms of causative mechanism and of evolutionary and philosophical significance. It is called *hypermutation*. It is the basis of affinity maturation and of a substantial increase in the total repertoire of B-cells. When mutation hits its target, a nucleotide in the complementarity-determining region, it can either remain silent with no effect on the amino acid expression (for instance, when the substitute radical is similar to the original) or be effective, and produce the loss or the substitution of one amino acid in the paratope. In this case, the effect produced in the antibody affinity could be a *decrease* (down to the disappearance of the antibody activity) or, more rarely, an *increase*. Fortunately, at the end of the day, it will be the rare positive change that will be selected, because it will overshadow all original paratopes of the clone in the race to secure the antigen's epitope.

There are some questions that hypermutation raises, which should be discussed here. Why is *the improved functionality* the *less probable* outcome of the mutation? Why are the two outcomes not equally probable as any random event should be? Well, here is why: one can imagine all possible paratopes specific to the antigen, as forming *a solid cone*, whose tip is the single *best fitting*

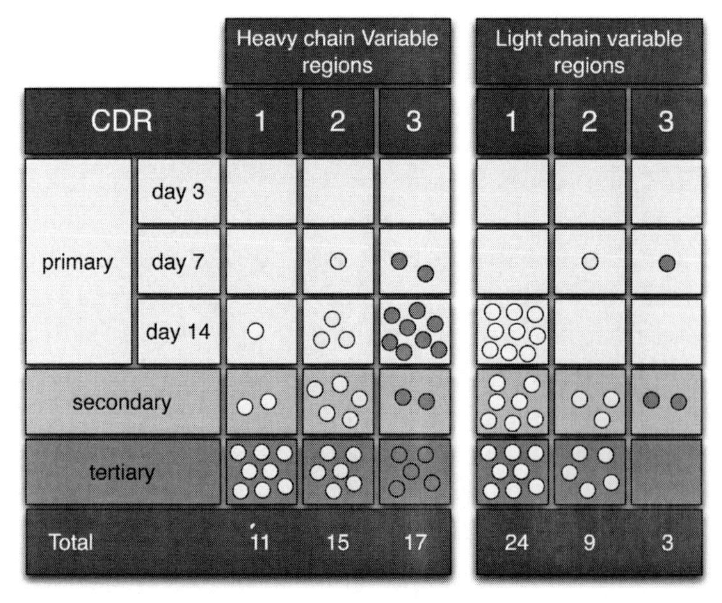

Figure 7. This figure shows the classic observation by Cesar Milstein that mutations appear in the CDR of the antibody, forming clones in late primary response and at higher rates during the secondary and tertiary responses. All three time frames are characterised by active cell division activity, especially the latter two. For this experiment, Milstein has used the technique of hybridoma that he has developed to produce monoclonal antibodies: antigen-challenged B-cells are hybridised with myeloma cells; immortalised cells are selected for antibody production, then grown *in vitro* and cloned from single cells and studied [8].

Colour image of this figure appears in the colour plate section at the end of the book.

paratope. One step down, say *degree two* in binding: they are more than one, because there are more ways to differ one step from the frontrunner, and that is the reason of the conic shape (it is a matter of combination, an alien would say). Even more structures will be found at –2 steps and still more at –3, and so on. If a clone is specific for the epitope, all its components will fit at a given level of the cone, corresponding to their affinity. Now, if one B-cell of the clone undergoes a somatic point mutation, the paratope will go either up or down one floor in the cone, and since there are fewer spots in the upper than in the lower floor, *Queen Probability*, her eyes bandaged, with her forefinger, indicates down.

1.20 Philosophy Storm ♞ ♜ ♛

Hypermutation helps clarifying the terms of the philosophical dispute in science which took place between 1930 and 1960, filling with excitement, the generation of researchers that were lucky to be active at the time. The question in everybody's mind was *specificity*, and how it was possible for a large but finite system to respond to all thinkable antigens, including those that were synthetic, i.e., man made, and did not exist in nature? Biology had been conquered by Darwinism during the previous century, and now—*anathema*!—theories were proposed for antibody formation where the antigen served as a template and *instructed* the host to produce a protein structure that was complementary to itself! In 1940, even the Nobel laureate, chemist Linus Pauling, thought that nascent protein chains would acquire the 'antibody shape and specificity' after synthesis, by molding themselves on the antigen molecule during the first encounter. "This made immunology the last bastion of Lamarckism", said Ruggero Ceppellini, repeating Luria's flowery prose about microbiology. Lamarck is famous for explaining (in 1809) the long necks of giraffes as the result of unrelenting piecemeal efforts by their ancestors to reach and feed on the leaves of tall trees; the association of Pauling with Lamarck gave him a philosophical stature, and produced a shock. Another shock was the bold contradiction of the hierarchy DNA>RNA>protein in the transmission of information (at that time RNA viruses had been known). It was with a deep sense of relief that witnesses saw, by 1950, that bastion begin to crumble. Darwin came back; biochemists rejoiced, having avoided the separation of structure and shape in proteins and the new, increasingly accepted theory was *genetic* and *selective*. The principal names associated with it were Macfarlane Burnet [9], Niels Jerne [10, 11] and many others. The antibody specificity was not *induced* by antigen but synthesised using codons randomly assembled within each maturing B-cell. The latter would bind only if the antibody receptor showed specific affinity. If so, the cell became a clone, ready to initiate, at the next signal, a massive production. Like so many events pertaining to evolution the antibody saga is impressive for its apparent waste of energy, cells and effort. However, nature may be blind, but not dumb. She does not want to throw away the work that goes into the construction of an antibody if it binds the antigen, only because its affinity is low. Thriftily, she applies hypermutation, a perfect Darwinian tool, where the antigen has furnished the energy for the binding (and probably the non-specific *danger signal* [12]). The change, hopefully an improvement, is stochastic, and the antigen has its final role as a selector. Since this process will be repeated, it strikes us as an admirable exercise of modulation, not so different in its end results from what the Lamarckians imagined, and also similar to the ideas of those who are always ready to accept new signs to hinting at *creationism*.

And what about the conformational effects in proteins? (see chapter 1.52). They are certainly not involved in dubious philosophies, but they look rather similar to induced biological information, as they pay no respects to the DNA>RNA>protein chain.

As soon as the induction versus selection battle ended with the latter's victory, a new war of theories erupted. This time, everybody involved accepted that antibody and T-cell receptors are coded by DNA, like any other protein chain. However, the germ line supporters proposed that the entire array of coding genesis was inherited, a product of the germ line percolated to the single lymphoid cells of the soma. On the other hand, those propounding somatic differentiation thought that the inherited information was as tiny as the code for a single antibody per individual! This DNA would duplicate furiously (adding mutations at each round) in the soma of the fetus; at birth, all the lymphoid cells would have a different code and specificity. The somatic differentiation seemed to win the dispute when it was shown that the germ line theory would require more than half of the genome space to provide coding for the antibodies, and therefore, it was impossible.

However, in 1976, the dispute was resolved in the field by a monumental work by Tonegawa and his group at the Basel Institute for Immunology [13]. It turned out that both theories were, in part, right. As already mentioned in 1.8 and 1.20, a significant fraction of the coding segments rearranged in the soma originates in the germ line. This means that it is inherited from the line that begins with the Lampreys. This also means that man is a citizen of the vertebrate family and probably enjoys immune capabilities and specificities selected for in the earliest of times but which are possibly helpful even today.

There is no firm data available on this point, but the hyper variable parts antibodies carry are witness to the unity of life on earth.

1.21 Suicide ♞ ♜

The wrong response can lead to suicide by self-aggression. The suicide weapon is autoimmunity: self and non-self do not belong to separate repertoires. The solution—a long and painful one—has emerged by natural selection at the level of organisms: negative for those who do not thwart auto-aggression and indirect damage, and positive, in the case of mutants that are able to control autoimmunity and the non-necessary responses. It is known, *a posteriori*, how these controls function: by implementing a moratorium to any specific response until 'independent information' is obtained that shows it to be safe to the body.

1.22 In the Beginning ♞ ♜ ♛

The transition from a state of rest to the frenetic activity leading to the duplication of all cell components and of the lymphoid cell itself, occurs any time an appropriate stimulus produces 'activation' (typically, the binding of paratope and the specific epitope), and is followed by cell division. To become 'activated' is a discrete step and entails the propensity to diffuse stimuli among the cellular society along with the capacity to kill and internalise invaders. It is the discrete transition from level 0 to level 1. It is the gate to survival at the cell and organism level. However, it is ambivalent, as it carries a shade of danger, a probability of suffering damage by the organism and its components. Controlling the distribution of 'activation' to the cells allows governing of the immune system. What came quick and easy in the beginning becomes the key focus of regulation. There are many aspects to this matter, and the discussion will continue for a couple of paragraphs.

1.23 The Linear Transmission of Activation ♜ ♝ ♛

The final adaptive immune system at rest (level 0) has no activated (B, T) effector cells (level 1); binding antigen, or contacting non-activated helper cells do not produce activation, and may sometimes cause death (this is the case of naïve Th, freshly graduated from the thymus). There are certainly many ways to break this state of apathy, but one is particularly informative and well documented. The way evolution chooses to defuse the danger of mounting unnecessary responses is to improve on intelligence. Only *certified* dangerous invaders will be met by response (a response is like a battle in the midst of a marketplace and should be only allowed in an emergency). This is an important point, considering the pains taken to control it. The stimuli that can *activate* the first line sensitive cells are chemokines which testify to the seriousness of a situation, and in particular to death among the organisms' cells: heat-shock proteins, necrotoxins, etc. Only these kind of signals can—now—break the pervasive apathy of the organism. Immunologists have known this for a long time before grasping the significance. But they have concocted the *adjuvants* (any of several mixtures of irritants, mimicking the presence of an infector) to trick their rabbits and have them produce all sorts of useless (to the rabbit) antibodies.

1.24 Recipe for Complete Freund Adjuvant

Add:

- one part of protein dissolved or suspended in saline (the protein is the antigen against which you want to produce an antibody);
- one part of vegetable oil;
- one part of mineral oil;
- A certain amount of heat-killed or dried mycobacterial bodies (for instance the Bacillus Calmette-Guerin, an attenuated strain of the mycobacterium of bovine tuberculosis).

Mix the components and emulsify them in a blend or by forcefully passing the liquids together through a large syringe until the aspect of a soft white homogeneous cream is reached.

Inject subcutaneously in mice, at the most, 1 ml distributed in several sites. (Alternatively, inject the entire dose, intra-peritoneal. Scale up for rabbits.)

Dr Jules T. Freund has been among the most famous and frequently cited names in immunology. The use of the *adjuvant* has made the production of antibody in laboratory animals reachable by anybody. It was finally realised that the failure to respond to antigen inoculation can be very revealing. And adjuvant—labelled by Janeway 'the little dirty trick of immunologists'—conceals that. In the 1970s, the theory of the 'two signals' became an instantly popular frame to explain the adjuvant action and the hesitations that must be won over to stimulate the immune response. Two signals are necessary, the first is the specific contact between the paratope and the invader's epitope, the second is a contact—or short range message—issued by a cooperating cell of the body: the dendritic cell that has contacted the antigen and verified its danger to the body, or an expressly activated macrophage. The second signal is also called *co-stimulatory signal* or *co-stimulatory molecule*. What is more important than its molecular identification, is its key position in the decision-making chain, where its presence will tilt the action from avoiding damage to making a life-saving response despite the harm to the organism.

1.25 An Antibody Is an Antibody (or Two?)

The antibodies are sensor-guided missiles of the immune system. Their existence was discovered by the Emil von Behring in the 1890s, and already in 1906, they were produced in horses and launched into patients with *diphtheria*; many were saved. It took much longer to understand the nature and the architecture of the little weapons. They turned out to be rather complicated, and they can be looked at from several views and at several depths. This means that no single picture is sufficient.

Looking at the left panel of Figure 8, the black bars represent the bonds that link the monomers and the dimers and keep the complex structure together. By breaking these linking bonds, four single chains with no antibody function result. Exposure to

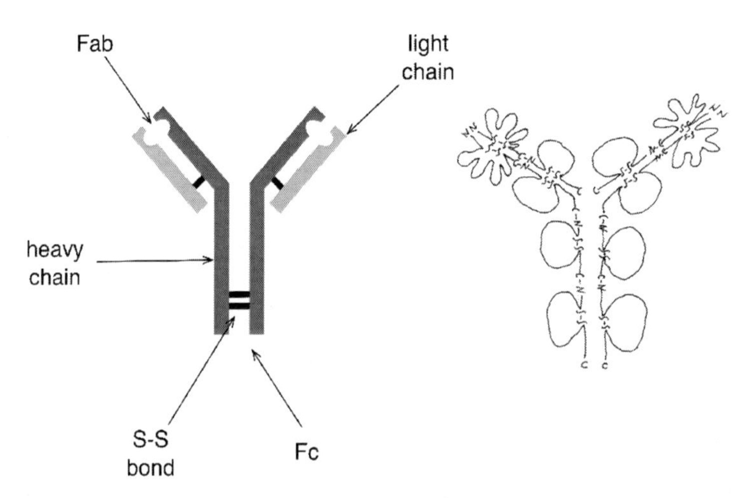

Figure 8. The left panel shows the structure of the antibody, as it was established around 1960 [14, 15]. It shows a molecule with a sedimentation coefficient equal to 7S consisting of two identical dimers, each formed of one heavy and one light chain. The right panel shows the way the antibody is an assemblage with a total of 16 strings, each consisting of around one hundred amino acids.

Colour image of this figure appears in the colour plate section at the end of the book.

papain, a selective proteolytic enzyme produced by the papaya fruit, breaks one peptic bond midway in the heavy chains disturbing neither the inter and intra-chain noncovalent bonds (S-S) nor the others; as a result, the molecule is divided into three fragments, two identical *Fabs* which retain one *paratope* with antigen binding function each, and one Fc, which is labeled 'c' for its tendency to crystallise spontaneously in the tube. The paratopes are formed by the folding together of the C-terminals of the heavy and light chains of each. The terminals are the 'variable parts' of the two chains, and consist of about one hundred amino acids whose sequence reflects the rearrangement—in the maturing B-cell—of three fragments of the *germ line code*. Thus, only antibodies derived by the same B-cell or clone are identical to each other: the astonishing consequence of diversity.

The right panel of Figure 8 considers the sub-structure of the heavy and light chains. The first is made of four of these units, joined together by peptic bonds, while the latter consists of two units. One essential characteristic of these immunoglobulin units (and of the evolutionary success of this extraordinary class) of proteins is the presence of two sulfur-bearing amino acids at positions 30 and 80 circa. The locking of the two SH groups into SS bolts confers the characteristic aspect of a structure made of repeated loops, typically conferring rigidity yet remaining elastic and robust. The antibody molecule consists of two ligated, identical heterodimers, each carrying one heavy and one light chain. Heavy and light chains are independently synthesised, and then assembled, in the same cell. In contrast to all other units, defined *constant*, the N terminal units of both H and L chains are *variable*, that is, their amino acid sequence is coded by RNA resulting from a somatic recombination of germ-line material that takes place in a single precursor cell and is shared only with the components of the clone resulting from its multiplication. The two variable H and L units face each other and delimit a three-dimensional space called *paratope*, where a specific epitope may bind. The two paratopes of the same antibody are identical, and all antibodies produced by the same clone are identical.

1.26 Scientific Myths: A God Interferes with Human Cognition ♞♝

Apollo, in love, endowed Cassandra, the daughter of King Priam of Troy, with the knowledge of the future, but later turned sour and condemned her prophecies to never be believed: Greek tragedies resulted.[3] In science a similar curse may hit truths that should be evident and easy to grasp. In 1916, Dr Freund had in his beaker the solution to the problem of immune regulation, but he did not understand it. Perhaps he considered his invention too modest to be meaningful. The same curse hit his colleagues and five successive generations of immunologists (each generation takes fifteen years to develop); they used the adjuvant because it allowed the production of all sorts of antibodies but did not understand the meaning of their action. They thought adding adjuvant was a simple encouragement to the response. Instead it was breaking one fundamental evolutionary principle: do not make unnecessary responses. No wonder so many results were misleading!

In 1963 Jacques Oudin, in his gothic laboratory on the third floor of the Institut Pasteur on Rue du Docteur Roux in Paris, obtained rabbit antibodies against the specific binding region of rabbit antibodies, the paratope [16]. This was a brilliant finding and he gave the name *Idiotype* to the epitope (the external shape) of the paratope, which is the antibody's receptor for antigen (Figure 8). At the same time, Henry Kunkel, at the Rockefeller Institute

[3] To be fair, Apollo (who toyed with Cassandra, as noted above) and Zeus (who punished Prometheus for smuggling fire to the Humans) were not alone in their anti-knowledge attitude. The monotheistic God of the Bible and the Koran is on record early, with the metaphor of the Tree of Knowledge, whose fruits man should not eat, and his alleged Representatives were active in the late Renaissance: the Inquisition sent Giordano Bruno to the stake in Campo dei Fiori in Rome in 1600; the Holy See forced Galileo to a shameful denial of his Science in 1633; and the elders of his synagogue excommunicated Spinoza in 1656.

in the Upper East Side of Manhattan, found that antibodies had antigenic characteristics related to their antibody specificity [17]. In the next ten years many researchers injected antigen into mouse #1 and produced antibody 1 (Ab1); then injected antibody 1 plus adjuvant into mouse #2, isolated Ab2, mixed it with adjuvant, injected into mouse #3, collected Ab3, and so on, continuing the series up to Ab7, and further. Interestingly, some Ab2s turned out to mimic the antigen and their use in place of vaccines was envisaged. An amazing development!

In 1974, the former fellow Franco Celada happened to be sitting in the dark wood lecture hall of the Institute Pasteur when Niels Jerne presented a revolutionary paradigm of the structure, function and control of the immune system based on the network of *idiotypes* [11]. The idea was fascinating. The anti-idiotype was called the *internal image*. It substituted the antigen, and all repertoires would be created *from inside* the system and evolve according to an internal logic, towards unknown directions or to no direction at all. Stability was provided by the simple, partly demonstrated theorem, that any anti-idiotype response blocks the production of the idiotype-carrying antibody by thwarting its clonal expansion. Mathematicians and physicists in the audience were crazed with enthusiasm. Others wondered why, but should have figured it out: they were aliens. An entire generation of immunologists was also deeply affected. Many of them were highly enthused. Jerne, who had already attained celebrity by founding the Basel Institute of Immunology, left the task of hammering down the details of the new theory to the well trained (and financed) cohort of young scientists that colonised the left bank of the Rhein.

Why then, in the 1990s, did the momentum of the network seem to falter? And how is it that in the 2000 edition of top textbooks, only one paragraph is dedicated to the idiotype network? Here is the sad truth: everybody that counted has uncritically accepted the adjuvant-assisted stimulations as a perfectly natural, albeit magnified, response. Instead, the adjuvant introduces signals that transform a protein that the system would have labelled 'not to be responded to' into an antigen against which the response is mandatory; the concoction fools the honest immune system and the researchers do not see that their theoretical construction is based on the consequences of a trick. What a blunder! It must have been caused by some aftershock of Cassandra's curse. If you try to obtain Ab2 by injecting mouse #2 with just Ab1, you obtain nothing because mouse #2 does not detect any danger in the injected protein and only a danger signal processed through the APC would convince it to change its mindset. It took quite some time for the reality to dawn on the sophisticated idiotype network people that they had based their admirable construction on responses that the immune system tries frantically to avoid as 'not necessary', by denying them the 'second signal'. They were forced onto the organism by a violent act, and therefore—in natural conditions—do not exist, in any effective form.

1.27 A Monument to Interdisciplinarity ♞

"We are in 1946, one year after Hiroshima and Nagasaki, which prompted the end of the war by causing the largest single tragedy of all wars! We are still busy assembling and storing nuclear bombs, but we are ignorant about the effects of radiation on living matter and the future of this planet!" This voice, with the heavy German accent that his colleagues called *wonderful*, is that of Alexander Hollaender, biophysicist, NIH scientist, delivering a speech to a government commission in Washington. The commissioners looked at each other and thought, "We did not know this guy was such an idealist... Well, nobody is perfect, he may still be able to handle our hot potatoes in Tennessee." Then they thought some more: the potatoes were really hot. The Manhattan project had ended, all good scientists were leaving in all directions to rebuild their academic careers, those who remained were mostly the medical staff which kept the plant workers reasonably out of trouble by monitoring the level of radiation, and who now found themselves at the end of the road ...

"Doctor Hollaender, won't you consider taking these employees, now at the Clinton Laboratories (one of the synonyms for the Manhattan project) and become their leader in the newly approved Oak Ridge National Laboratory?" Hollaender was shocked by the shallowness of ideas the commission was exhibiting. He said nothing, and kept staring at them through his glasses, with an expression half benign and half dumb. The commissioners were puzzled, near the point of giving up, but said, "Would you at least be willing to take a look at the place and write us a report?"

He boarded the first military plane that serviced the high security area at the NE corner of Tennessee. Two months later, he was back with a complete report. Franco Celada, born in 1931, would have paid to see the faces of the commissioners as they were reading it, but they were in Washington and he was beginning high school in Milan. Hollaender had designed a unique, interdisciplinary institute around the interaction of radiation with living matter. The report is in the archives. Here is the juice with a hint of added drama.

Radiation means alpha, beta, gamma, neutrons, protons, X-rays and cosmic rays. Each of them acts differently, and no target tissue or cell is altered the same way or has the same susceptibility. It was necessary to study cytology and genetics at many levels, down to the nucleic acids (even if a reasonable idea of DNA was still ten years away in the future).

"It is necessary to study many different organisms: Bacterial and Drosophila genetics are a mine for gene function, but to study genetic radio-damage that may affect humans, we have to use mammals. For instance, is the genetic damage produced by a given radiation dose delivered in one minute or spread over ten years the same? Probably not, on account of self-repair mechanisms, but we must find out. Here the flies are no use, since they live two or three days. I will study this in mice. They have a lifespan of three years, which is OK but not optimal. We will have to extrapolate from three to ten or twenty years. This is possible if the data are statistically solid. Therefore we must be prepared to use them in high numbers. Mice are fortunately easy to breed. We will plan a colony of at least 50,000. Pathologist and immunologists will look at acute damage to cells and tissues, but all combined and considering 100 animals per experiment and 2 experiments per researcher per week, another 50,000 per year must be programmed. We will invest a lot on immunologists. In academia they have forever played second fiddle to microbiologists and pathologists and they feel ready to take big strides. Also, there is a hint that the immune system may be the crucial hot spot of radiobiology; at the same time, a target for severe damage and an obstacle to remedy, if the remedy should turn out to include cell and tissue replacement. We will verify this point!"

The report proposed the initial deployment of six independent sections, Biochemistry, Cytogenetics, Physical-Chemistry Physiology, Experimental Radiology Zoology, and a group dedicated to collaborations, intra and extramural. All sections would be housed together in an industrial building constructed to purify U235 by mass spectrometry (but eventually not used since an alternative and simpler methodology proved satisfactory). The site was called Y-12 (see Figure 11). Concentrating the labs in one site would generate interactions and encourage collaborations between labs and disciplines. Initially, the external collaborations mentioned were with the radio-physicists of the Clinton Laboratories and—with higher emphasis—with the National Institutes of Health in Bethesda, and with seven Universities, including Princeton, Johns Hopkins and Tennessee's own Vanderbilt.

The commissioners realised that they had aroused the genie: no options now of putting him back in the bottle.

"And what about the staff?"

"You see that on the next page: 70."

"Is this figure a 5-year goal?"

Figure 9. This is Alexander Hollaender posing in front of a magnificent Capogrossi and other pieces of his collection. Not being super-rich, he adopted as collector a strategy very similar to the one he used to hire his scientists. In both cases, he was trying to find his objects of desire two or three years before they reached notoriety (and, in case 1, became too expensive or, in case 2 were hired by Harvard). He had a network of colleague friends and a network of shrewd art dealers in most European centres. He used to visit both networks during his periodical travels to Europe. In the years after the War there was a boom of modern painters in Italy. They figured generously in Hollaender's collection, which had also found in the visiting immunologist Gino Doria, a keen spirit and informal advisor on Italian modern art. In the late 1950s, there was a prominence of young Italian post-docs at the Biology Division. What caused this correlation is anybody's guess.

Hollaender smiled under his moustache; he had won. "No, that is the initial figure; 70 by 1947. After that, the reasonable annual growth rate will be 10%. By 1960, staff will be 200."

Then he explained: "Since I count on interdisciplinarity, I need them all in Oak Ridge in a hurry. The more young, brilliant scientists we have there, the more of the same quality will be willing to come, and not only from the USA. Living in a desert will be no obstacle; it will mean adventure. We have learned a lot about population dynamics during the war and particularly at Oak Ridge. I don't need to explain to expert managers like you, the concept of *critical mass*."

The project, as outlined by Hollaender, was approved without the usual delays, the self-standing institute called Biology Division was inaugurated in 1947, and despite all imaginable difficulties, took off and grew along the guidelines of the initial report. The predicted inter-disciplinary collaborations did take place, and Hollaender rounded up his educational objectives by adding two sections to the Division: Mathematics and Editorial. Celada can speak from his direct experience, which began in 1959: when a researcher planned an experiment he would visit a mathematician, to discuss, e.g., how many observations would be necessary to publish his paper with a reasonable hope of not being disputed. To him, statistician M.A. Kastenbaum said, "In research you are never *sure to be right*. You may *feel* confident, but in fact you make a *bet*. You bet that what you see in the mice of your experiment applies to all mice on earth; and what you agree to put at stake in this bet is the most precious thing for a scientist: *your name*. What I can do to help you is to measure the *probability* that the conclusion, based on your sample, might turn out to be caused by a fluctuation of numbers instead of corresponding to a universal truth. Is this probability one out of ten (0.1)? One out of 20 (0.05)? One in hundred (0.001)? Publishing data at 0.1 is clearly foolish; however, the decision is not mine, but the author's, that is, yours. What kind of man are you? Do you take risks in your life? Are you a gambler at heart? If you are not, there is only one way: more work, more time, more mice, to make the adverse probability shrink, becoming smaller and smaller."

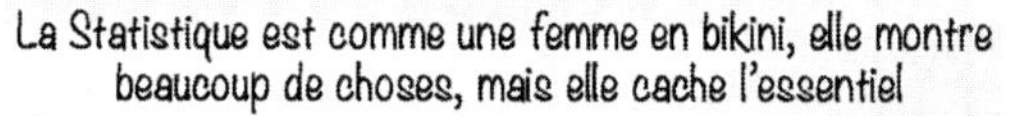

Figure 10. The authors do not really know if statistics resemble a girl in a bikini, but the irresistible power of the French joke, and the enthusiasm of the mathematicians of the Oak Ridge National Laboratory, have persuaded them to publish this picture—contemporary with the facts narrated—of Brigitte Bardot, convinced that it is the nicest bikini they have ever seen, or dreamed of. As for Dr Kastenbaum, the Director of Biometric and Statistics, when he first heard the definition from a young research fellow, who had picked it up in Paris two years before, he erupted into hearty laughter, and asked for the text to be painted over his office door. Before leaving in 1961, Celada passed by to greet his statistician friend, who showed him the almost finished inscription. Maybe it is still there. As for Brigitte, she wrote a lovely letter authorizing the use of this image, and encouraging our research in the hope that modelling will replace the use of animals in biological experiments.

Kastenbaum was also giving formal lectures that had a great attendance. So were the writers of the Editorial office, whose courses were advertised *"for visitors"*, but were attended by many Native Americans who profited even more than us. The goal here was to improve the scientific prose: clarity, no reverse passive phrase and, of course, no prohibition against writing 'I'

Figure 11. Oak Ridge National Laboratories, the Biology Division, in a photo of 1976, kindly provided by the Oak Ridge ORNL Library. The building, identified as Y12, had been planned as part of the Manhattan Project in the 1940s, and should have hosted an industrial scale ion-spectrograph for the purpose of isolating U335, the fuel of atomic bombs. When another method of Uranium purification proved successful, the works at Y12 were discontinued, and the site was occupied by the ORNL Biology Division in the mid-1950s. Some preliminary tests with Uranium must have already been made, since during the first years the biologists were not allowed to access some parts of the large building on account of the radioactive contamination level.

Colour image of this figure appears in the colour plate section at the end of the book.

in a scientific paper if that made the expression more natural and easy to understand. The speakers had a great time reading excerpts of real manuscripts before and after the cure that they had suggested. Celada's first paper in Oak Ridge (the first he wrote in English) took ten re-writings, but was also the last one to require such a heroic effort.

1.28 Scientific Myths 2: Apollo Roams about the Hills of Tennessee ♜

In the 1950s the Biology Division began to grow, and within a few years it became a beehive of projects. Especially interesting were those regarding fly and mouse genetics and those produced by the pathologists, who experimented on rodents but aimed to apply their eventual results to treat radiation exposure of humans. Rapidly, they found that mice and rats exposed to more than 900 RAD would die in 5 days, but could be saved if infused with bone marrow cells from the same species, and even cross-species, e.g., infusing rat cells into mice. This was an important result, and ready to be applied in cases of accidental exposure of workers in the atomic plants (minor accidents, regrettably, did happen). Voluble Apollo must have helped the pathologists to a point (as divine protector of the arts, he wanted to see Hollaender's collection of modern paintings, and was walking in the hills because he shared with him the hobby of hunting fossils of ancient forests). But then, like he had done with Cassandra, he turned away and the resulting spell prevented the researchers from understanding what they had uncovered at the bench. Instead, a furious discussion went on for years; they were sure the effect of the bone marrow was due to some *humoral* factor, but they disagreed about the molecular nature of the factor. In case the experiments of bone marrow infusion had been only performed by mouse-to-mouse infusion, the hypothesis of a real 'take' of the donor cells and their taking over host haematopoiesis would have certainly been considered. What made it unthinkable was the presence of the life-saving effect also when the marrow donor was a rat: another species!

1.29 Plain Reasoning Breaks the Spell

Dan Lindsley—30, handsome, smart, pole-vaulter and butterfly collector by hobby—was among the drosophila geneticists of the Biology Division. He had no reason to be interested in the rodents, and had not met Apollo. Thus, no help, but no spell either. At lunch, and at seminars, he heard the discussions of the pathologist and his eyebrows became increasingly arched. He remembered a demonstration in college where the professor had shown that rats had genetically determined blood groups, and their red cells could be typed by using agglutinins, just as in human blood typing. The man who made the demonstration was a cult mammalian geneticist, Ray Owen, from CalTech. Dan admired him and was sure he could ask him to send a few of his rats already typed (*in research nobody is at all alone, and nobody is at all naïve*). He approached two of the younger pathologists: "Are you able to extract bone marrow from rats?" "Can you inject cells into other rats?" "Will you be able to do the final blood typing of the recipients?" The answers were, "Yes, yes, yes…" Dan shouted: "Then, what are we waiting for?" The experiment was not discussed further. Lethally irradiated rats were given group-different bone marrow. As expected, they did not die, even if they became anaemic and weak. After twenty days, they were all recovered and their blood was of the donor type, proving the functional implantation of the donor bone marrow cells. In November 1955 Lindsley, Odell and Tausche published a paper in the Proceedings of the Society of Experimental Biology and Medicine [18]. Hollaender was bursting with joy. Dan Lindsley was invited to speak in several meetings and to the New York Academy of Science. He savoured this success, then retuned to his flies, as if nothing had happened. At 87, he says that he has still two or three drosophila experiments to complete. The results described broke the spell, but did not formally demonstrate inter-species transplantation. This lack was soon filled in 1957 by Takashi Makinodan and Nazareth Gengozian of the Immunology Group on the 4th floor of the same building. The god, Apollo, was nowhere to be found, but Ray Owen, who had indirectly inspired Lindsley and was spending a year at ORNL, was co-author in their definitive publication about inter-species transplantation. When in 1959 Celada won an NIH Post-Doctoral Fellowship, his Director Ruggero Ceppellini said, "That's a hell of a place. You must go there!" It was one of the rare things the two agreed on.

1.30 Of Wolves, Dogs and Shepherds ♞ ♜ ♝

A *quasi-Aesopian* metaphor can be used to depict some aspects of the foreign infection at the onset of the immune response. In this metaphor the sheep are epithelial cells, the dog is a dendritic cell, and the wolf an infector. The dog sleeps, unnoticed, among the flock of sleeping sheep (see panel 1 of Figure 12). In panel 2, the wolf trespasses and attacks one sheep. Roars, noise, saliva and blood are sprayed around and reach the dog. In panel 3, the dog wakes up and becomes excited (activated). He engages in a gruesome battle with the wolf. His mouth full of bloody wolf fur, the dog starts a long run to reach the shepherd's hut (panel 4). The shepherd hears barking, sees the dog's evidence, understands what has happened and becomes incensed with rage (activated). In panel 5 the shepherd's offspring (a clone of activated Th cells) contacts hunters in the hut. The hunters become enraged and, fully armed (activated), run down the path, ready to kill any wolf they will find (panel 6). More impersonations: the flock is an epithelial organ, the path is the lymph vessel, the hut is the lymph node, the shepherd is a T helper cell and the hunters are cytotoxic T-cells. A partial variation of this story has the enraged shepherd meet (and activate) a technician (the B-cell) able to manufacture missiles radar-guided to the strain of wolves that committed the aggression. This represents the humoral response, and the missiles are the antibodies that will land specifically on the aggressive wolves.

The actions in the cartoon are derived from the data gathered by Antonio Lanzavecchia, and take place in a rural landscape where the sheep are epithelial cells attacked by bacteria (the wolf) and defended by the dendritic cell (the dog). During its process of activation, the Dc undergoes period of rest, increased mobility and change of direction, along the path of the lymph vessel, to the lymph node, where it meets and exchanges biochemical information with a specific T-cell (the shepherd), which is in turn activated via lymphokines and becomes a clone whose components descend through the same vessel to attack the infectors.

Figure 12. The wolf, the sheep and the shepherd dog in a metaphor of the immune response. The description of the events is to be found in the text.

1.31 How to Block Anti-self ♞ ♜ ♛ ♚

Anti self-responses are the other, major, danger that motivate the wave of controls that will modify the specific responses. In addition, the selected solution is blocking activation, this time at the end of the line, at the level of the effectors, B-cell and T cytotoxic cell. Since it is expected that only non-trigger happy organisms have survived the first self-aggression wave, any effector response needs specific T-help. An obligatory enrolment in the organ thymus of all bone marrow born T lymphocytes is the selected method to single out, and eliminate, most of the prospective T helper cells specific for self. These killers work via apoptosis when the exposure to self results in a high affinity interaction. The iconic picture that results is that of a B-cell specific for an ambiguous epitope, which resides on a self-antigen. The B-cell binds it, internalises it and exposes significant peptides on appropriate receptors. Then it waits. And waits. But the Th will never arrive, because it does not exist anymore. And the B-cell dies.

1.32 Controlling Self-damage Needs Structures: The Lymphoid Organs ♞ ♜

Organs dedicated to implement controls are the thymus (which negatively selects anti-self and positively selects anti-nonself cells), and lymph nodes, which facilitate the encounters of antigen-presenting cells with T-cells and of T-cells with B-cells. The maze of meeting places and of lymph-vessels results in a mass of lymphoid tissue that weighs more than two kilograms in man, while the complexity of the system grows in parallel through evolution. As control grows in efficiency, activation becomes more difficult to achieve. Lymphocytes that were trigger-happy become hesitant. While awaiting the results of the control, many legitimate responses are lost. By the time auto-aggression and immunopathology are thwarted, the anti-invader responses are slow and weak. The history of the immune system progresses by *waves and undulations*.

1.33 Time of Strengthening ♞ ♜

When all possible controls are finally in place, the danger of self-damage is at a low level, but that is forcedly true for the efficiency of immune defences also, and a low tide in defence efficiency must have followed (unfortunately, nothing is known about the defences of the dinosaurs. If this was known, they could make a nice example in either case: for being too controlled, and thus having become extinct for insufficient defence, or for being not sufficiently controlled, and aggressive against themselves). After the low tide, the phase of reconstructing efficiency begins. Infection and invasion remain the main selective forces; targeted organisms, in order to survive, must exhibit improvements, for instance, ways to circumvent some controls without re-opening the door to danger. It seems an impossible task, but Evolution's ingenuity should not be underestimated. One of the key improvements is to build up memory. Every step is governed by chance and selection, but this one has the ring of Niccolò Machiavelli, subtle and persuasive, advising his Prince on policy, "Listen Monsignor, we (the organism) have increased the blocks on all responses, and that weakens us. We should exempt from blocks, all responses that are safe!"

"Caro Niccolò, which responses are by definition safe?"

"Those that already ran in the past, and have not damaged us. Therefore, let us uphold them! Let us give them a free pass! Let us prolong their life, and let them have all the help they need without waiting days and weeks in line, like new responders must do! And after the response, let us keep them around, ready for quick action like Navy seals."

1.34 Maturity

With these adjustments and exceptions in favour of these adjustments, the system reaches maturity. It can be proud; it has introduced a revolutionary weapon that had never been tested before. The daring experiment was on the brink of backfiring.

Thus, its actions have been, and are, severely controlled. The relative weaknesses and slowness derived from multiple regulations have been absorbed and compensated. Humans have learned the basic rules to take advantage from memory, even if they sometimes they forget what the cows have to do with vaccines. It is time to ask the burning question: Is it true that the immune system knows what it is doing?

1.35 A Cognitive System? A Floating Brain? ♞ ♜ ♝

Umberto Eco was the right person to consult with about this. The four-man committee (Sercarz, Celada, Mitchison and Tada) invite him to the hills above the charming town of Lucca; the date is September 9, 1986. He has just arrived with assistants and students from the University of Bologna; the car, dark blue, was bought with royalties earned by 'The Name of the Rose' in USA. "We see that the lymphocytes and other immunocytes send and receive messages to each other, and the immune response is triggered or blocked, depending on the nature of their contacts. We need your authority in *Semiotics* to decide whether the immune system can be considered able to understand what is best for the body. In other words, is it a sort of brain? A mind?"

Figure 13. The group talking with Umberto Eco: (from left to right) Umberto Eco, Eli Sercarz, Franco Celada and Masaru Taniguchi. Photo courtesy of Rabyn Blake.

Colour image of this figure appears in the colour plate section at the end of the book.

Umberto Eco scratches his head, then, characteristically scratches his beard while making a grimace. He thinks that in order to answer this query he must subject the immune system, or even the single immunocyte, to the same kind of test that he has adopted for the mind: that is, to present them with an inherently *ambiguous* signal, and study their reaction. He explains to us, "If ambiguous, a signal has two possible meanings. The human mind considers and develops both and stores them as *hypotheses*, and then freezes. It will choose between them *only if and when* independent information is acquired. Then he asks, "Can you suggest an immunologic ambiguous signal that we could use in a cognition test?" Of course! The first that comes to mind is the case of a *self*-molecule that exhibits some epitopes identical to those of an alien invader (a case with definite probability). What happens when the B-cell has bound this antigen, perfectly fitting in its paratope? Launching the response would be self-aggression! But the B-cell is aware of such mistakes and does nothing, except waiting, until the appropriate Th cell arrives with news that the antigen is foreign, or does not arrive. Eco looks first sceptical, and then he becomes interested. "The B-cell hesitation is unexpected! This is different from the kicking you observe when you tap the kneecap with a rubber hammer. Or when you push a button and the bell starts ringing. It seems to me that we are witnessing an unconscious way to *understand* the nature of the stimulus, and to make the right response under the present circumstances, even if the cell should die in the process. The immune system is an unconscious mind. Primitive perhaps, but a mind."

1.36 Cellular Automaton Makes Sense ♞

No sooner has the meeting at Lucca finished, than the rush starts because, after all the intellectual excitement, there is a lot of work that somebody has to do. It was a pity that the IBM man Philip Seiden, the astrophysicist, could not come. But back in New York Celada tells him what Francisco Varela, a Brazilian working at the École Polytechnique in Paris, had presented. The title was really catchy: Structural Coupling and the Origin of Meaning in a Simple Cellular Automaton [19]. In room 1611 on the top floor of the Hospital for Joint Diseases, Philip Seiden is pleasantly surprised. Cellular Automata, invented by Von Neumann in 1935 [20] and developed by Stephen Wolfram [21], are his speciality and he has used a similar discrete system to model hydrogen molecules aggregating in the interstellar space to form a galaxy. Philip explains to Franco: "Von Neumann started a Mathematics using geometric (1, 2 or 3D) entities instead of numbers, and the possibilities are astounding. The most obvious is the 2D variety, which take the shape of a surface divided into discrete squares, like a chessboard, or like a tiled roof. By the way the adjective 'cellular' has nothing biological about it, but derives from the Latin word 'cella', a tiny room. Each site has at least two alternative states, like empty vs. occupied, or black vs. white, or even showing a number of alternative colours, or shades."

"You mean a chessboard can change from black to white and vice-versa?"

"Sure, and do not worry: these boards are not used for the game of chess. The point is that the change is triggered by conditions that may happen in the sites immediately adjacent to the one we are observing."

"You mean the change-producing conditions are known? Which are they?"

"No, no! It is the modeller that is entitled to write the rule, a set of simple laws that characterises his particular automaton! A different rule means a different automaton and a different result. Another responsibility of the modeller is to decide the aspect of the game at t=0, but he can also choose the density of one of the states and simply rely on chance for the aspect at t=0. The dynamics begins with the application of the rules to the situation of t=0. The changed situation will appear at t=1, and the rule will again be applied, with certainly different effects."

"Is this the Game of Life? And what has life to do with it?"

"Franco, hold your horses! The Game of Life is one particular 2D binary example, whose rules, dictated by John Conway in the early fifties, proved able to generate a great variety of shapes and seeds: combination of several same-state sizes able to reproduce themselves periodically."

He shows me, on screen, the most amusing automaton, the *Game of Life*. This construction by John Conway 'occupied the mind' of a number of East Coast university students in the 1970s. It agrees with the formal definition [22] of a Cellular Automaton as a discrete lattice of sites, which deterministically evolves in discrete time units, where each site can take a finite number of states. In the Game of Life, the rules for the evolution of a site are exclusively based on its state and that of the eight adjacent sites, and they are simple: (i) a black (i.e., live) cell with fewer than two live neighbors dies, as if due to under-population; (ii) a live cell with two or three live neighbors lives on to the next generation; (iii) a live cell with more than three live neighbors dies, as if because of overcrowding; (iv) a white (i.e., dead) cell with exactly three live neighbors becomes a live cell, as if by reproduction. The game begins by randomly throwing any number of black 'cells' into a white space, and all changes are deterministic. The East Coast students have tried millions of initial random combinations as 'seeds' on their rudimentary computers and have discovered a biological trait of the game: a number of the seeds happen to have a distinct character, or 'meaning'. In Conway's collection of seeds, the frontrunner consists of only five black dots occupying nearby cells. This particular combination, when the rules are applied, changes shape at each time step, yet it remains made by just five cells. After five iterations it returns to the original shape but has moved one step right and one step down in space! The resourceful *glider* has a motor (Figure 14E). There are many initial combinations with interesting capabilities. One is actually found *on commission*: Conway offers 50 dollars to any student who discovers *the glider cannon*, a combination of black cells that would, in time generate a glider. He is happy to pay the price. Another find—much easier—is the *glider killer*; an object that,

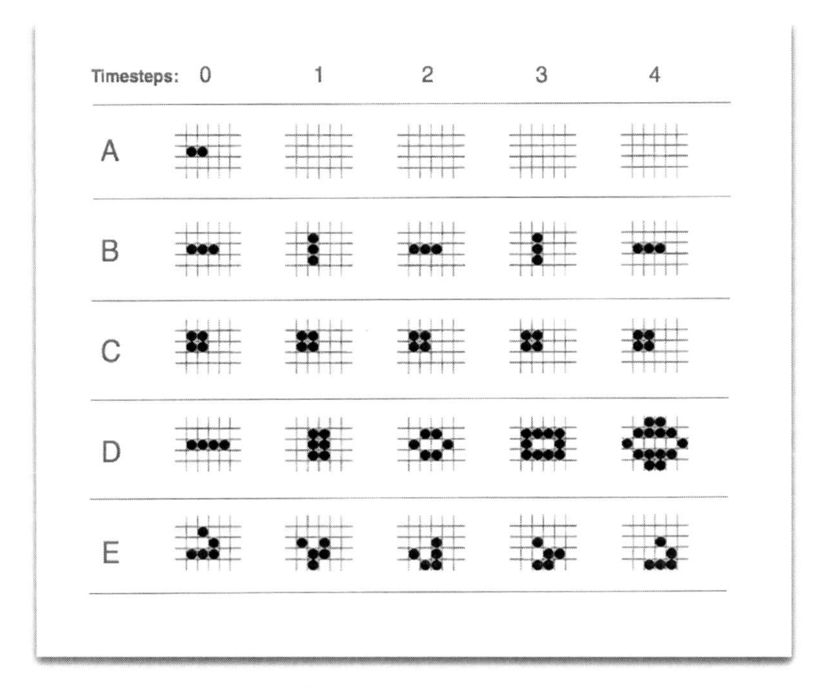

Figure 14. Examples of patterns' evolution in a Cellular Automaton defined by Conway's rules [22]. In row A, the *double spot* disappears at time step 1. In row B, a sequence of 3 spots (*the Blinker*) alternates orientation with period 1. In row C, a square patterns remains stable forever. In row D, a sequence of 4 spots changes shape and size at every step. In row E, a characteristic pattern of 5 spots (the Glider) keeps its size but changes patterns at every time step, returning to the first shape with a period 5, but moves one step down-right.

positioned on a collision course, will destroy the glider. The killer is on a suicide mission; no trace whatsoever remains after the collision (thus it looks more like *kamikaze*).

The reader who is wondering why the Game was labelled "of Life" might note the undeniable and almost disturbing assonances with "Life" in the popping up of "cellular" seeds, and the one-in-billions random sequence of atoms that, hit by a spark, happens to learn duplication.

1.37 Memory Is In the Numbers ♞ ♜ ♝

B-cells, when primed like the one described above, expand as clones, while thousands of similar cells do not, and die. These may have failed to bind the antigen because, a) their receptor does not grant the same level of affinity (that is, of probability of binding), or, b) having bound the antigen they may have failed, by hazard, to encounter a Th cell. Since all events in biology are probabilistic, it must be accepted that probability breaks down when small numbers are concerned. If two cells can count on a certain level of probability of binding (thanks to their mutual polar or hydrophobic attractions), how is this translated when the discrete meeting materialises? It will break down to a binary decision: yes or no, 1 or 0. One can use anthropomorphic examples, but nobody can influence the 0/1 decision. The happiness of a cell becoming a clone is palpable. Does it follow that the cells of the immune system are influenced in their behaviour (such as engaging cooperation with other cells) by their individual interest, and that behaviours like cell cooperation are established with this motivation? Perhaps. But only after the fact, natural selection will uphold or reject what happened at the cellular level. In health, the wellbeing of the organism is the interest to which smaller units must pay respect. But it is not an order imposed from above; the organism comes first, simply because if it dies, or suffers, all components die or suffer.

1.38 The Power Rests with Whoever Issues Regulatory Signals ♞

The appearance of specific repertoires of receptors at the onset of the adaptive response is the dowry carried by the newly formed lymphoid cells. It became their trademark and was engineered though technologically advanced immunoglobulin class *glycoproteins*. Judging from the current situation (in Italian, *col senno di poi*), this combination must have exhibited early evolutionary advantages. Since the life conditions in the primordial oceans do not suggest high stakes, the advantage of a specific immunity may have taken the shape of help to the social life of early vertebrates (see the lampreys in section 1.5) rather than that of ammunition in their war against bacteria. As a matter of fact, the early immune system at this stage may have become a source of danger to self, and this explains the drastic imposition of controls (by the patient selection of the less trigger-happy members of the species). Different controls are applied to different dangers. Let us examine and distinguish between the two varieties of self-damage produced by launching a response: *specific attack* (by cross-reaction or by mistake) and *non-specific* damage from inflammation, suffered by the body *as a bystander*. Both are serious but their controls follow separate paths.

Direct anti-self is controlled by making sure that no specific responses mounted are disturbing self. The Th cell is charged with this certification: if it binds to a class-2 receptor on the effector cell, it means that the effector is not anti-self, the signal is *GO*, and the planned response is launched. To deliver a STOP signal there is a logical shortcut; since the self-binding Th were rightly suspected of aggression, they have already been eliminated in their thymus selection. The STOP is delivered by *lack of Th binding*. The cells waiting for the messenger will continue to wait, and then eventually die.

The bystander damage (aka, friendly fire) is 'preferably' avoided. However, in the presence of real dangers, as in case of a pathogenic infection, the no-action policy is lifted; the body accepts the price of going to war. The lifting signal is issued by the dendritic cell, which is informed by and informs the T-cells. The nature of the danger signal is highly stimulating and pro-inflammatory. Both signals alerting the dendritic are mimicked by—guess who?—Dr Jules Freund in his adjuvant. Anti control A is the one acting on anti-self responses and anti control B is the one taking care of the bystander damages.

It is interesting to note that A and B are the expressions of two different evolutionary steps and thus of two cognitive modes. The former is manned by T-cells, expressing the adaptive response philosophy, while the latter is carried out by the spearhead of innate defences, the dendritic cell. Immunologists, and, apparently, also Semiologists, are in admiration of the elegance of these controls and break-controls. To better understand their nature, immunologists try to learn to command these signals, in both A and B cases and for both the GO and the STOP sign. The way to revert the STOP anti unnecessary responses (AUR) was found in the adjuvant (see section 1.24). The way to block the GO signal to AUR is to use all sorts of anti-inflammatory drugs, from Aspirin to Cortisone (here the pharmaceutical industry has uncovered a gold mine). The control of anti-self response has generated brainstorms in only one direction: how to revert or substitute the signal GO to anti-self, in order to allow the friendly and life-saving transplantation of organs and tissues. Unfortunately for the immunologists, it has remained only a brainstorm up to now, while effective tools have been developed in the hands of the producers of immunosuppressant drugs (which block clonal expansion of anti-graft responses), or in the hands of growers of monoclonal antibodies anti HLA determinants (with more moderate success).

1.39 The Seed Is Prepared During the Primary Response ♜

Let us return to the accomplished responder: the B-cell that has bound, internalised and presented antigen fragments on its class-2 MHC receptors. If it encounters a specific T helper (which has been previously activated in an encounter with a dendritic cell), then it undergoes activation, which leads to cell division and cell maturation. The clone emerging begins two strategic actions: the production and secretion of antibodies, and the seeding of memory. In order to perform these tasks, more intra-clonal changes take place: some members of the clone community continue to multiply for some days, up to 4 times a day. Then they stop and mature into production machines, and lose the capacity to proliferate further. These are called Plasma Cells (or plasma B-cells, PLB); their antibody production goes on for months and covers the needs of the primary response. Some initial clone members are kept by the body for future use; they do not participate in the exciting primary antibody synthesis but conserve the genetic makeup and the capability to reproduce, if needed, and re-enact the response that has just been made.

The seeds of memory have acquired a longer life, they look like young B-cells, and they preserve, in their DNA, the precious recombined sequence that was put together by the ancestor B-cell described at the top of the paragraph. It is not known how many 'identical' seeds are produced at any primary response. Certainly, many. And since the primary B-cells were probably more than one, of similar specificity but not identical (remember the cross-reactivity mentioned earlier?), the resulting memory cells are several groups at each response.

1.40 The Challenge ♞

It was Influenza. The response was successful. Time passed. Years went by. Then again fever, again flu, and again the same strain. Immediately the B memory cells, worked up by cytokines, are ready to bind. Let us look at this one: it internalises viral material and displays 9 to 14 amino acid peptide segments on class-2 receptor molecules. Naturally, memory cells also need the Th help to start. Do you remember how difficult and uncertain it had been for the precursor of this cell to find a specific Th? Well, it is completely different for this B memory cell. She is sure to get help, no waiting at all, and the reason is that during the primary response, the meeting of B and Th was celebrated by a squirt of cytokine, which induced both cells to each produce a clone in a hurry. Even after years have passed, numerous effector clone members and Th clone members are around and amenable to a specific meeting like before.

The memory cell has to take another step before secreting antibodies: it has to become a new clone one more time. The cell goes through a number of duplications of its genetic material, including copying the codes of the antibodies. In each gene duplication, some error may occur in the hyper-mutable zones. The process is called 'hypermutation'. This phenomenon has been discussed in sections 1.8 and 1.19 and illustrated by Figure

7. The final outcome is an increase in affinity with important positive consequences, among which is the enlargement of the repertoire of the individual, and the second positive selection of the memory cells for not only the present engagement but also for the future.

1.41 Competition In the Minefield of Cross-Reactions ♞ ♝ ♛

This, as it appears now, is a dynamic society of cells and clones, motivated by competition for antigens and for signalling molecules. Special receptors crave to bind them via the noncovalent forces that energise all protein–protein interactions. At stake is survival; the competitors belong to the same clone or, with more dramatic effects, to different clones with different histories, such as naïve cells and memory cells. At each appearance of an antigen, a race starts, and not all runners will be pleased. The memory clones, enhanced by the power of multiplicity, have the best chance, especially if re-challenged by an identical antigenic molecule. Otherwise, if the second antigen is different, it all depends on *how much* different. If it is *similar*, it can be assumed that not all clones in the memory will bind it. Those that do will be advantaged, in a measure proportional to the affinity they display. Those that don't will be obliged to leave space for the new responders; their cells will commit *apoptosis*. This entire procedure is called *attrition* (see 1.48). During the history of an individual, there is as a tendency toward selection of cross-reacting memory clones. It is easy to see the advantage of this. However, it seems to entail a trimming of specificities that goes in a direction opposite to ever-expanding diversity: a nuance of regression.

1.42 The Compensated Immune System Wants to Understand ♞ ♝

The immune system that emerges is very complex and flexible. Its primary goal, before starting any defence, is to understand if and what—in the inside and outside environment—is threatening. This, in order to decide whether to mount an appropriate response or not to respond at all. The time it takes to gather information, labelled *hesitation,* is the trademark of cognitive processes. In this interval the immune system asks: How big? How numerous? Was it seen before? Is it dangerous? Did it kill innocent cells? Is it part of the body? Is it visible from the outside? Do any of these cells hide something suspicious? When all these questions (and many more) are answered, the hesitation is replaced by action or actions. The response will be launched on both channels (humoral and cellular) and will be a fast one if the attack is recurring and memory can be utilised, or a slow one—time being needed to build up populations, and to pass more controls—if it is a brand new one.

1.43 The Two Immune Systems ♞ ♜ ♛

In 1998, Franco Celada and Philip Seiden travelled to San Diego and said, "One of the most interesting things about the immune system is that it is not one, but two: therefore we must model both, at the same time." The duplicity has been known for a century and a half. The very first descriptions by Emil von Behring of anti-toxin antibodies, and the observation of phagocytes destroying microbes in wound tissue, revealed defence mechanisms belonging to different worlds. The two forerunners, the solid upper class German and bear-shaped, wild, bearded Russian at the *Institut Pasteur,* their location in the most developed nations of Europe always in competition or to war against each other, helped a lot in igniting grandiloquent and emotional scientific discussions. The history of immunology has since shown that humoral and cellular are two arms of the same body, different in strategy and in mode of detecting and attacking invaders, activated almost always together by all emergencies. They don't really collaborate, but they lend parts to each other as good tinkerers do. The deployment of different defences and the decision of which to use in any particular occasion enhances enormously the efficiency of the immune system. But there is also *competition* between the two branches for resources and notably for antigens. The *two systems impression* is produced by the plasticity of the defences, which evolved their own philosophies to be able to cover the diversity of menaces from foreign aggressors. Suffice it to consider the attacks from outside and the attacks from within that can be suffered by host epithelial cells, as epitomised by a bacterial and a viral infection. The branches have developed quite different strategies, specially aimed at attackers using different strategies.

In the case of bacteria, their aggression is aimed at damaging or destroying chemically, by toxins diffused or by actual membrane contact, the cellular epithelium; by doing this they are obliged to expose their own surface, which is covered with sites that the body's antibodies will be able to reach. The binding antibody tags the aggressor's membrane and serves as an anchor for deadly enzymatic strikes by soluble chain-reacting molecules (called *complement* for reasons too difficult to explain). This looks straightforward if compared with the mission of the cellular branch.

1.44 The Cellular Branch Confronts a Virus Infection ♜

The virus is an incomplete organism that can only reproduce itself by penetrating into the cell and borrowing the cell's ribosomes. When it enters the body, there is a narrow time window, of seconds or minutes, during which it is vulnerable to antibodies, then it penetrates the cell membrane and becomes invisible to the humoral branch of the immune system. Inside, it finds both sanctuary and structures allowing it to reproduce itself in a hundred thousand copies. The infected cell and the immune system (its cellular branch) act as if they know that there is no way to cure the cell, and that the only hope for the organism to survive is to kill every infected cell with its smuggled contents of viral proteins (or virus particles in different stages of assembling). In order to succeed, this operation must be done quickly, but how can the system identify a single infected cell in a large population of cells? Simple, by asking it to confess. This is done by implementing an executive order: every cell of every organ of the body must surrender, in real time, samples of any product of their ribosomes. The samples, in the form of 9 to 14 amino acid chains, will be sent to special receptors called *class-1*. Here they will be mounted on the receptor and appended outside, where they will be inspected by specialised patrolling T-cells, the cytotoxic T lymphocytes (CTL).

There is no 'general inspector cell'. Every mature CTL participates in the search. Lymph nodes are bursting with circulating, stroking, contacting CTL that probe all class-1 loaded receptors. Every mature CTL has its individual (T) cell receptor (TCR), which has already been approved by the thymus as able to bind to 'a foreign epitope presented in association with a self receptor' and will only detect whether that class-1 is presenting *the* epitope for which she is specific. In the rare event that this happens, it will say so, by binding at high energy, and killing the epithelial cell, destroying its secret viral contents at any stage of assembling. Considering the number of epithelial cells in the whole body, and of the different viruses that may penetrate them, it may seem to be like searching for a needle in a haystack. This impression is justified. What makes this heroic enterprise successful at a level that is acceptable by evolution is the size of the receptor repertoire and, therefore, the large number of the different T-cells available and in use, *that implement* specificity.

And the healthy, innocent *epithelial cells* which are always the vast majority? To the inspecting T-cell, they show only peptides of self class-1, to which the T-cell responds by developing a low energy attraction, sufficient for a passing stroke of approval, not for an interaction. That is reassuring in a point of history where everyone is controlled, and hopefully approved of, in a silent way. Better than other periods, like those populated by STASI, KGB, CIA, SS3, MI5, etc. Or is the silence a cover-up? The signature of the new Big Brother?

1.45 How to Model The Cellular and The Humoral Together ♞

1997. The IMMSIM simulator began extending its capacities in order to model, at the same time, both the cellular and the humoral response (with results to be shown in San Diego to attendees of the modest-sounding 1998 IEEE International Conference on Systems, Man and Cybernetics). It was a bold move because immunologists had rarely looked at the two defences together as being mutually interactive. The reason for this was, that with the advent of new tools, immunologists were busy going deeper and deeper into their own chosen fields, collecting more and more data on smaller and smaller details, becoming more and more specialised, and forgetting to take into account whether (or deciding to ignore if) the two branches interacted at all. On the contrary, the discrete IMMSIM that would have had some difficulty at the sub-cellular level was completely at home when asked to follow all the steps of the engagement of the two responses, to run them in the same space and the same time and in the presence of the same virus infecting one epithelial cell population. And this is not all. The discrete model, as Seiden engineered it, was totally transparent and totally manageable, whereas the mouse or any *in vivo* culture was not. Immediately, the size of this advantage was clear. Not only was it possible to vary doses of the infecting virus, its speed, capacity of penetration, etc., one could run the same infection in the same virtual organism with *only* the cellular, *only* the humoral, or with *both* branches working. The results could then be compared and the contribution or the disturbance introduced by each deduced quantitatively. The biologists' dream, realised.

1.46 Body and Soul, Actions and Philosophy ♞ ♜ ♝

It is worthwhile to print some excerpts of the 1998 paper [23], not only because of the initial effort to model 'the entire' immune system at the same time, but because it was in this paper that the IMMSIM-team declared their objective to embed in its code, even if in a minimalistic way, the acts and the principles that inform it, as a *cognitive system*. Here is the excerpt from the introduction of the paper [23], in the hope that its declarations may be somehow useful to the aliens of today:

> An interesting view of the immune system has developed, over several decades, from that of a rather mechanistic response-to-stimulus machine to that of a cognitive system capable, not only of building a specific response, but of making the decision whether to enact it or not. It achieves this by using information independent of the stimulus per se, and following a behavioural code evolved during phylogenesis and engineered in ontogenesis. From what results from the accumulated findings, the connections among lymphoid cells look more similar to those among neurons, and the recognition/response follows rules similar to those defined by semiology for the human mind [24].

Breakthroughs

Two breakthroughs stand out in the recent history of immunology in this respect: the concepts of cell-cell cooperation (Th-B [5], Th-APC [25, 26]) and of Tc-target interaction through peptide presentation on autologous MHC as defined by Zinkernagel and Doherty [27]. They are the basis for understanding both 'cellular' and 'humoral' responses, and they reveal a natural philosophy for survival underlying the immune system, whose tenets or principles summarise eons of trials and errors.

Principles

The first principle is: "Be prepared for any possible foreign invader." This is mainly achieved by building an almost infinite repertoire of receptor specificities.

The second principle is: "Do not inflict damage to self; directly or indirectly." These two principles are in a 'Catch-22' relation since, by enhancing the repertoire of specificities and the power of the effectors, there is an automatic increase in probability of autoimmune aggression by sheer cross-reaction. The evolutionary solution of this paradox was in the separation of effector power from decisional power, assigning them to different cell lineages (on the effector side, Tc and B; on the regulatory side, Th1 and Th2) that recognise separate sets of specificities. As a result, interaction and cooperation between cells takes place at all critical points of the immune response. This type of cooperation allows, for example, the B-cell repertoire and recognition capacity to remain virtually complete, as long as the T helper cells from which the B will receive the 'go' signal are carefully selected in the thymus and will only recognise 'non-self' antigenic determinants.

This arrangement goes a long way towards reaching an effective compromise and avoiding the conflict between the two principles above. However, there are loopholes threatening to nullify the control provisions. One is again cross-reactivity between self and non-self, now at the level of the T-cells: it has been calculated that to avoid it completely, the thymic negative selection should eliminate most of the T specificities that are essential in responding to foreign invaders.

Another loophole is created by intermolecular help (IMH), a phenomenon occurring any time an unexpected noncovalent bond occurs between a self-molecule and a foreign molecule (e.g., antigen-antibody, CD4-gp120, etc.). Thus, if the self molecule happens to be recognised by a B receptor (B-cells have an almost complete repertoire), the foreign antigen will be co-processed and presented, and consequently will provide to the B-cell, the T help that would have been lacking for the self protein (thanks to the thymic negative selection). Numerous cases of IMH-caused autoimmune responses have been detected recently [3, 28], including rheumatoid factor formation, and anti-

CD4 antibodies in a high proportion of AIDS patients. This scenario will occur in most infections, as the foreign invader establishes a noncovalent link to any host component.

1.47 The Selective Modeller ♞

It was a considerably momentous decision to model the quite different dynamics of the cellular and the humoral responses; that of running the simulations together was a great step towards biological truth, but did not require additional efforts. The architecture and all the key cells of the system were already there, some commands on the computer array just had to be flipped from 'sleep' to 'go'. Figure 15 shows the steps that were implemented, and which can serve as blueprints to reproducing or upgrading the model. Before reaching the stage of drawing algorithms on the coding paper, the modeller must understand what happens, either in the mice and cell cultures on the bench of his immunologist friend, or according to *the state of the art consensus*. This tuning of minds is not written and really agreed upon officially, and does not exist for all processes at all times: it must be *interpreted* by the immunologist of the team (and his interpretation always means *simplified and cut down to the bone*), according to his intuition, taste and philosophy. Complexity tends always to increase in biology, because nature is a tinkerer, and the result is endless duplications in processes and mechanisms. There may exist ten ways to go from A to B—and this constitutes an evolutionary advantage *if something should happen*—but only one or two may really be used. The model cuts complexity by implementing only one, the most efficient. It can also consider two or more steps if they are in direct causal relation with each other. In the twin Figure 15, effects like activation of cells upon contact or of proximity to other cells can be noticed. In reality it is known that lymphokines are involved. If modellers want to study the lymphokine deed, they should model it as a molecule, with precise parameters, functions and rules. Alternatively, if they want to focus on the logic of activation, they can, in good conscience, introduce the rule that a given cell

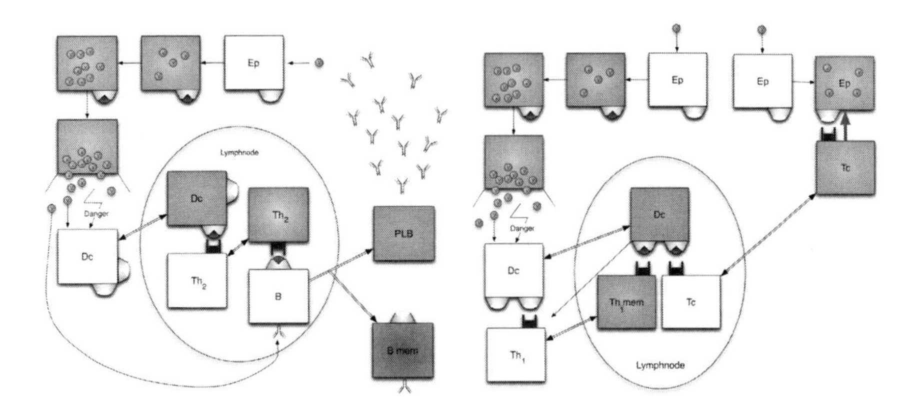

Figure 15. *Left panel*: The humoral response. Starting from the top and proceeding counter clockwise, the virus enters an epithelial cell Ep and begins to multiply. In succession, Ep displays part of the virus (V) on its MHC class-1 then lyses, liberating virus particles. Some of these are picked up by a resting macrophage or dendritic cell (Dc), which also perceives a damage or danger signal. Thus, it transits to an aroused state (light blue shaded Dc), and moves towards a regional lymph node. Here it displays a viral peptide on its class-1 MHC molecule, and if a virgin Th2 makes specific contact, it is activated and will be able to, in turn, activate a B-cell. The latter multiplies and differentiates into antibody-secreting plasma cells (PLB) and B memory cells (B mem). The antibodies are able to inactivate live viruses but are harmless for virus nested inside target cells.

Right panel: The cellular response. The triggering is identical to the humoral up to the activation of the APC, which then binds, on its receptors, both an effector and a helper T-cell. Both are activated by the Dc, and the 'help' reaches the effector (Tc) thanks to the proximity of the two T-cells. Consequently the effector multiplies and its offspring are ready to bind the class-1 receptors of any infected epithelial cell and kill it, thereby destroying also the incomplete virus particles present in the assembly line. There is consensus that the 'help' can still reach the effector even in case of a temporal discrepancy at the time of the binding of the T-cells to the APC.

Colour image of this figure appears in the colour plate section at the end of the book.

causes the automatic activation of its target simply when the distance required by the lymphokine action is reached. A large simplification will be achieved, and the model will not suffer for lack of realism.

1.48 Attrition and Cross-reaction ♜ ♝

Attrition is good onomatopoetic word because it imitates the sound or the sensation of objects rubbing against each other while being squeezed into a space too small to contain them. The word is used to describe what happens when a population of cells, such as a group of a memory clones stored in a lymph node, is hit by an antigen challenge. One imagines the fervent activity when the memory CTLs, which secure the antigen, then expose class-2 receptors, are met by specific helper T-cells and begin to multiply, all in the time lapse of a few minutes to one hour. Immediately, the problem of space arises. The volume of stimulated clones can increase ten-fold, occupying all spaces in a lymph node, which is dilated and can be felt protruding and tender under the skin. What is called *passive attrition*—a sheer limitation of volume—will finally cause apoptosis and block the further expansion of the memory response. But there is also an *active attrition*, which takes place in the first phases of the stimulation, and was discovered by Liisa Selin at UMass, the University of Massachusetts in Worcester [29].

There is a characteristic dynamics of a live virus challenging an organism with preformed memory. If the new infecting virus is antigenically different but not totally foreign, some of the memory clones will cross-react, and some will not. The class-2 presented epitope will stimulate the former, and as part of the chemokine storm there is secretion of interferon (IFN-γ). Interferon will act locally and kill memory cells, probably by both responding and not responding. It is secreted early, acts quickly and makes space for the newly multiplying cells. It is the flywheel for proliferation. The end result is that cross-reacting clones are positively selected every time antigen hits them.

2002 AD. The modellers of New York, on the 14th floor of the Hospital for Joint Diseases, have just returned from the inaugural meeting of an interdisciplinary grant with Liisa Selin. They like attrition, and decide that IMMSIM could act as an illustrator, carefully incorporating the events of passive and of active attrition. Dario Ghersi and Claudia Calcagno, fresh MDs from the Medical School of Genoa, but mathematicians at heart, build a virtual mouse, endow it with memory, expose it to virus, and challenge it. Suddenly by an act of prestidigitation the virtual mouse has become three: one sports active attrition, one has passive attrition, one—probably the most realistic—has both varieties together. All the rest in them is identical, and easy to compare, in the total openness of their immune system as simulated by IMMSIM (see Figure 16).

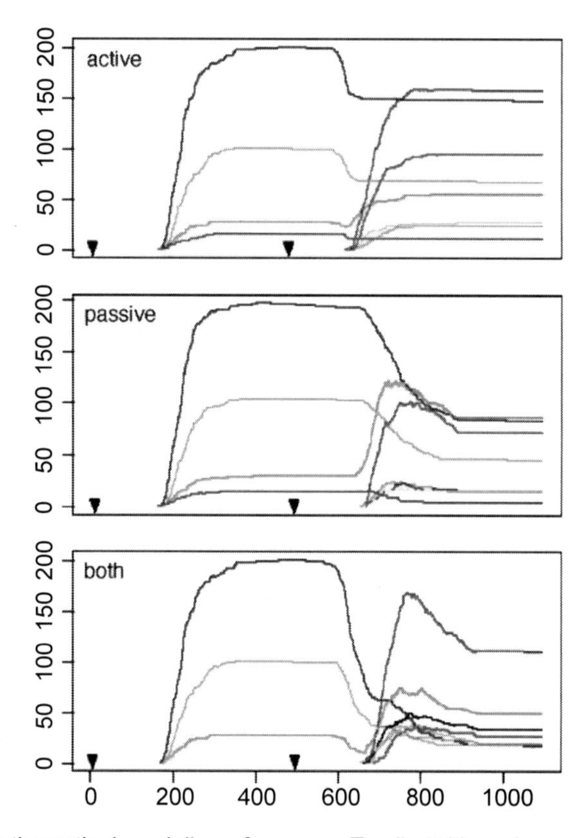

Figure 16. Mathematical modeling of memory T-cell attrition, demonstrating active, passive, or both mechanisms: Tc memory clone dynamics during the primary and secondary response to cross-reacting viruses, under three scenarios of attrition simulated by IMMSIM. Abscissa: time steps. Ordinate: Tc memory clone cell counts. Arrows at time step 1 and 500 indicate the time of inoculum of V1 (70 particles) and V2 (120 particles), respectively. V1 clearance is complete at time step 200 and V2 clearance at time step 700. Reprinted with permission from [29].

Colour image of this figure appears in the colour plate section at the end of the book.

1.49 Memory Is Strong, Sometimes Too Strong ♛ ♚

Memory cells have several advantages over naïve cells. One is their number, since many cells form each clone. They are long-lived and have preferential access to the lymphoid organs. They are easy to activate and their response is produced in minutes or hours, one magnitude less than unprepared cells, which need time to procure antigen and to build up the number of effector cells. This explains why, in the presence of a homologous memory, no naïve primary response materialises: the memory response has eliminated the virus and there is no antigen left to stimulate the naïve (that is the competition). The critical case is when the antigen does not correspond perfectly to the specificity of memory, i.e., it is cross-reacting [30]. Considering the mutation rate of viruses, this is more the rule than an exception. At intermediate antigenic distance, memory cells have low affinity, while some of the naïve cells have high affinity. But here also, the memory response is disproportionately favoured. Their mass (that is, the number of identical cells) will, in part, compensate for the low probability of binding. The speed of the memory response again tends to outplay the high-affinity naïve cell. Thus, the range of memory extends further than what is justified by its paratopes. An oligarchy is being installed. This is an asset as long as the response clears the infection. But, if the process goes one step too far (as it eventually will), memory will become a burden.

1.50 IMMSIM Reveals MaN (Memory anti Naïve) as a Cause of Aging ♛

To understand both causes and mechanisms of the senescence of the immune system one should not search for a dramatic pathogenic event striking in mid-age, but rather detect and probe among the common, repetitive events under one's daily observation, asking whether their consequences could accumulate during a lifetime and eventually build up to a no-return situation. While studying the commonest act, the

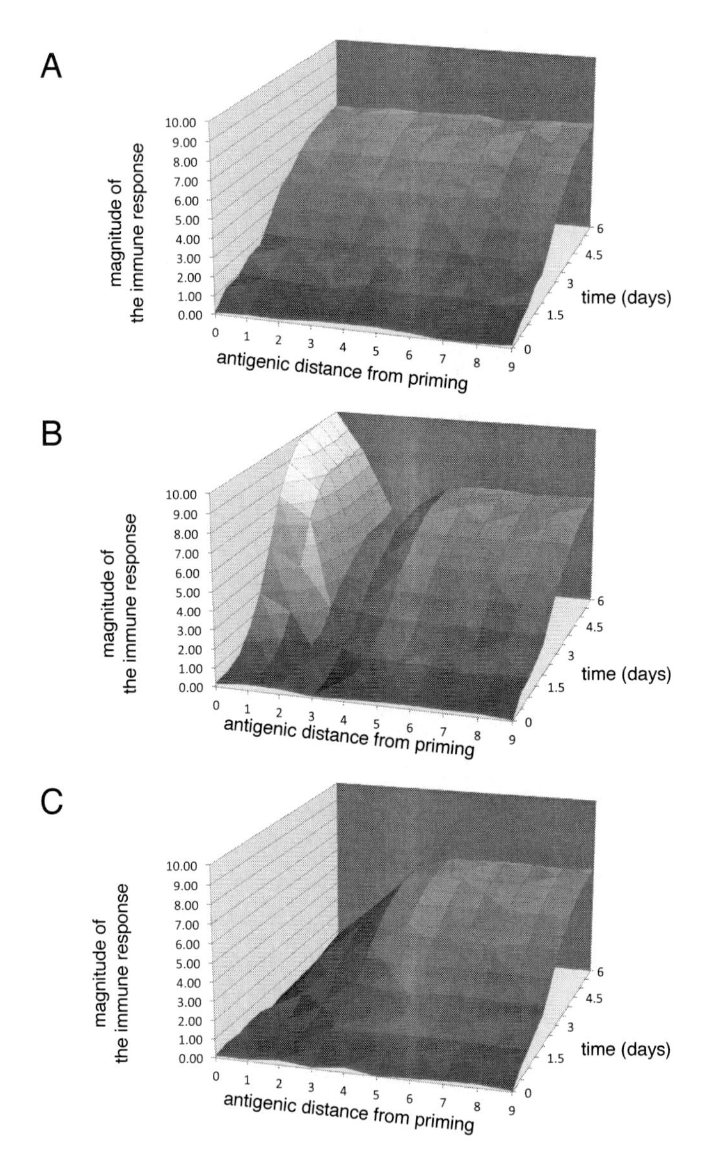

Figure 17. Experiment *in silico* demonstrating the MaN (Memory anti Naïve) phenomenon [31]. In panel A, naïve T-cells develop a primary response against ten virtual viruses differing from each other by one bit (out of 16) in antigen specificity: results of the ten runs, labeled with the antigen distance from 0 to 9, are shown side by side in the 3D representation. Each curve shows the kinetics of the cytotoxic response. Thanks to the large repertoire available, the naïve responses to the different viruses are not significantly different. In panel B, the cell population contains not only naïve cells but also memory cells primed with virus '0'. The 3D pattern shows the secondary

Figure 17. contd....

interaction between effector and pathogen—in a world of mutable prey trying to escape from specific predator defences—the homologous, perfect match must be rare indeed, while cross-reaction becomes the rule. A corollary follows: since the effectors in the cross-reaction have been previously primed by another antigen, they are, by definition, memory cells. The large number of clone members at hand (built up in preceding actions), as well as their speed of deployment, assures their predominance. Thus, immunological memory can compensate for its suboptimal affinity and remain competitive and, up to a point, lifesaving. This allows vaccination to be considered the best weapon against viral infection, wide in scope and covering long periods, provided the invader's mutation rate is not overwhelming (as in HIV). Thus memory extends its coverage, and earns public admiration, but its dominance is based on false pretenses: muscles instead of quality, cohorts of cells instead of highest affinity.

The mature immune system is now walking through a minefield. One historic observation [32] should have alerted immunologists about the growing gap between cross-reacting memory and infecting antigen; unfortunately, a victim of its 'catchy' name (Original Antigenic Sin), it was archived as an interesting but isolated fluke. The eye-opener came in the form of a systematic simulation in the IMMSIM computer model. The results both reproduced the antigenic sin and explained its mechanism and consequences: memory clones not only

Figure 17. contd.

response developing in column 0 and the decreasing cross-reaction in columns 1, 2 and 3. Note the higher speed and magnitude of the memory response. The primary response is completely wiped out in column 1, and severely thwarted in columns 2 and 3. Only from column 4 on, is the primary response undisturbed. The MaN effect is enhanced by the memory clone size and speed of growth, which together outcompete the naïve cells. In panel C, the same data of the experiment in panel B are plotted, but the memory response is artificially cancelled to better illustrate the 'hole' caused by memory on the primary response.

Colour image of this figure appears in the colour plate section at the end of the book.

dominate (as expected) the response to homologous and near-distance cross-reacting antigenic stimuli, they also manage to preempt naïve responses by using the large cell numbers as a mass action to defeat the few high-affinity new cells. Another advantage of memory's large numbers is the easy encounter between specific effector and helper cells, which is a risky and time-consuming step for primary responders. Thwarting naïve responses may have no immediate negative effect but constitutes a discrete and progressive loss of future diversity, a step in the direction opposite to the evolutionary expansion of repertoires that characterised the advent of the adaptive immune system in early vertebrates. MaN—the 'Memory anti Naïve' conflict shown by IMMSIM—emerges as an important, self-sustained causative event of senescence which, by creating and establishing an oligarchy of memory clones, eventually causes rigidity, imprecision and impotence of the adaptive system, bringing about most symptoms of hampered responses in the elderly and the frequency of immuno-pathology.

1.51 A Course Leading to Aging of the Immune System ♚

An oligarchy is formed, which tends to further increase its advantages. The life saving glory of vaccination carries a dark side. The diversity of the repertoires is compressed. The case of the *original antigenic* sin (see section 1.50) is brought up in the 1970s [32]. In 2009-10, IMMSIM predicts memory-anti-naïve suppression [31]. Lethality of the H1N1 pandemic of 2009 is traced to low-affinity immune complexes produced by cross-reacting humoral memory responses that cause lung immunopathology and death [33]. Low-affinity immune complexes are suspected to cause immunopathology because they are unstable; they hold the antigen for a while and then they may release it, causing re-stimulation, inflammation and cytokine storms. A group of Argentinian researchers working at Vanderbilt in Tennessee, led by Fernando Polack, stumbled on this finding by observing an unexpected age distribution of the H1N1 victims: middle-

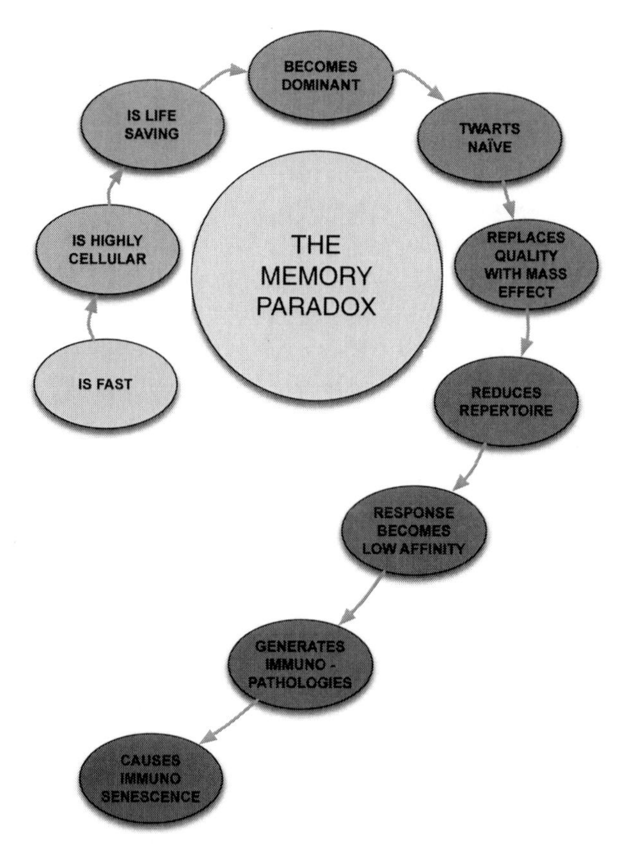

Figure 18. The attributes of immunological memory in colour-coded balloons according to positive (green) and negative (red) effects on the immune system.

Colour image of this figure appears in the colour plate section at the end of the book.

aged patients had fared worst and died more than the old and the very young [33]. They were able to explain this fluke when they found that many of the middle group had been infected by a *cross-reacting* strain about ten years before; a case of *antigenic sin* at its worst and saddest. The young were able to mount an effective primary response, while the elder patients had survived a previous H1N1 epidemic, thirty years earlier. This is an unexpected, welcome validation of IMMSIM's prediction. Moreover, it shows that MaN can become a serious threat.

1.52 A New Challenge for Modellers: Conformation

Cellular automata, or as they are called today, agent-based models, are efficient in simulating cellular and cell population dynamics, and this has been amply shown and recorded in literature. However, in principle, the discrete models can be usefully applied to all functions where the statistics of small numbers applies. The latter includes many molecular and even sub-molecular systems.

The field of protein conformation has many crossing points with immunology and its development seems particularly appealing to modellers. Thanks to the novelty of the effort and the expected boom in this field, this section may be of use.

In the three-dimensional space, the positioning of a protein and its parts is determined by the primary structure (the sequence of amino acids), but also by the noncovalent forces exerted within the secondary, tertiary and quaternary structure, that is, by the complex and changing micro environmental repulsions and attractions. Because of the environmental influence, there exists more than one conformation that a protein may assume, and these shapes correspond to low, free energy pits. At any condition, one conformation is more stable than the others, and enjoys highest probability and represents the normal shape and function of the molecule; but temperature, electric fields, other local forces could change it. All sorts of events may induce, switch on and off, modulate any function and any process, as evolution knows well and curious researchers are just discovering.

It should be kept in mind that conformation is augmenting any repertoire of capacities in biology and is obviously utilised by nature to command, signal and regulate. This is increasingly clear from the finding that conformation is, for large proteins, a social function, like spoken language for humans; the proximity and/or contact between diverse proteins is not neutral, it always alters something in the local environment, and may—and often does—favour an alternative conformation. That is, it operates a switch. This kind of behaviour can be spotted in systems as different as the regulation of the *lactose operon* in *Escherichia coli* and the activation of complement in immune haemolysis. Induced conformation as a serial domino cascade has been recently identified as the astonishing pathogenic event in *prion* diseases.

1.53 Protein Conformation and The Immune System

What opened the possible application of conformation in immunology is a historic truth: antibodies can recognise conformation. A rabbit is injected with a protein, say, Tetanus Toxin (TT). One millilitre of the resulting antiserum is sampled and the molar amount of Toxin that it can bind in antigen excess is measured (this requires the use of radioactive labelling). Then TT is incubated in saturated urea. This treatment does not disturb the peptic bonds of the amino acid chain but annihilates all other noncovalent forces around the primary chain, thereby destroying conformation. The treated antigen is used to repeat the measure of antigen binding. The result will show less than half the binding of the first test. This description is based on a classic experiment that demonstrates the existence of a relevant proportion of antibodies that bind to conformation-dependent epitopes, instead of linear epitopes. Of course, the antibody does not know it, but the atomic groups that properly fill its paratope belong to three or four amino acids that are not adjacent to one another, but belong to different sections of the antigen's amino acid chain, and only conformation brings them together and allows the Ag-Ab binding.

This antibody can detect the conformational epitope, and it can be used for this purpose but, like an electron shot by the physicist, it may alter the measured object. The antibody that recognises a conformational epitope influences the conformation of the antigen, usually by stabilising it, following the universal rule of keeping free energy at a minimum. However what the antibody sees may not perfectly satisfy the internal memory stored in its paratope; for instance, the antigen may have suffered a mutation, or degradation. In actual conditions, the binding would be *at low affinity*, as in the case of a *cross-reaction*. This could be considered to be the rule, but cases have been observed where the induction of a sizeable conformational change of the antigen molecule would bring about a reconstruction of the *perfect epitope*. In the present case, a small, refreshing, natural miracle takes place: the antibody tries hard to induce a conformational change in the antigen, and if the energetic equilibria are favourable, it will succeed. This means that the Ab-Ag bond will become optimal, and that other functions of the antigen molecule that a genetic insult had disturbed will be repaired! There are many observations that fit this description.

Figure 19. The figure to the left is a metaphor of an amino acid string that constitutes a protein. It has a shape reminding one of a musical sign. If the three intersections of the string are on the same plane, there are three points where amino acids interact with a noncovalent bond, and that fixes the shape and meaning of the figure. There is a way to eliminate the covalent forces, and that is treatment with concentrated urea. The result is shown on the right, where the same amino acid sequence has lost both its shape and its meaning. This change in the figure serves also as a metaphor of Antigenicity of the protein. One can imagine that a number of sequences of three or four amino acids act as linear epitopes. They will be equally well recognised by specific antibodies in the two versions, before or after the treatment with urea. On the contrary, there are up to three sites at the crossing points that may act as epitopes: these will consist of three or four amino acids belonging to two different segments of the string. They are called 'conformation-dependent' and will be recognised by their specific antibodies in the version to the left (before urea), but not in the version to the right (after urea treatment). The present figure illustrates conformational epitopes dependent upon the tertiary protein structure. In polymeric proteins, epitopes can consist of amino acids belonging to different strings altogether. These are called 'quaternary structure-dependent'.

Perhaps the most spectacular example goes under the name of *antibody-mediated activation of a mutant enzyme.*

1.54 Antibody-Mediated Activation of a Mutant Enzyme

Escherichia coli Beta galactosidase (Beta Gal for short), the product of the *lactose operon,* is the enzyme that allows the breakdown of dimeric lactose into more manageable *glucose* and *fructose* monomers. Humans don't have this vital tool and depend upon gracious *escherichias* to do the work as part of their intestinal flora duties. The enzyme is composed of four identical monomers. Of these, two joined head-to-head form a dimer.

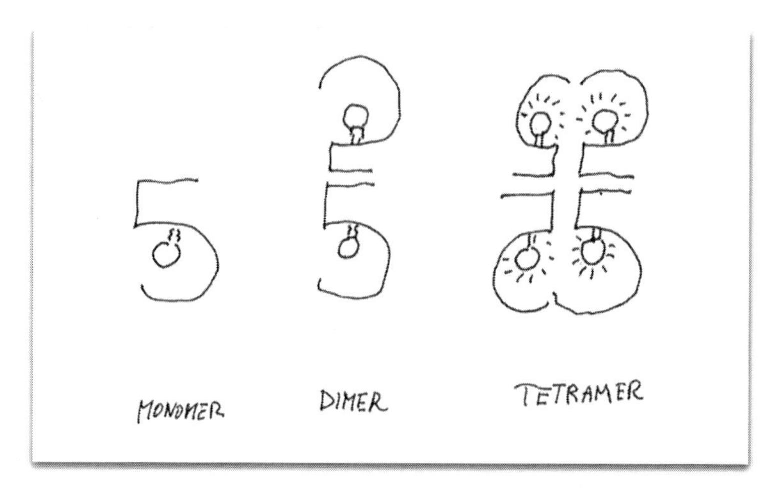

Figure 20. Arbitrary diagrams of Beta galactosidase monomer, dimer, and tetramer. This scheme was proposed years before the X-ray diffraction analysis solved the structure, but satisfies most of the shape and functional characteristics of the molecule. Two wild type monomers form a dimer and two dimers combine to form a tetramer, but it is only the latter that exhibits d-2 symmetry and has enzyme activity. The monomer chain contains 1214 amino acids and shows its C-terminus at NE and its N-terminus at SW.

Two dimers joined by their monomers' N-terminals form the functional enzyme. Its four catalytic sites are positioned like the opposed cylinders of a Porsche boxer engine. Naturally, Beta Gal is one of the fastest and most robust enzymes, and like in the Porsche, it will function only when the entire and intact tetramer exist. It happens that an induced mutation, changing one amino acid (out of total 1021) in the N-terminal region of a monomer, will completely impede the enzymatic activity. It turned out that the mutated amino acid was vital for the conformational architecture of the molecule. Could some 'patch' cure the damage? This was a quite crazy question, but Rotman and Celada asked it anyway [34] in July 1967, in Providence, RI and the answer was, 'Yes'. Antiserum, anti wild type Beta Gal, was able to produce a substantial amount of catalytic activity, with an impressive rise of three orders of magnitude! Not all antibodies in the polyclonal serum were able to do this, but of the fraction of the 'conforming' ones, even the purifier monovalent Fab fragment was sufficient to flip one enzyme molecule. The mutated Beta Gal formed solid dimers, but the resulting tetramers were unstable. The presence of specific 'quaternary' conformational antibodies would either add an external bond to the defective tetramer or induce a conformational transition towards the functional 'dimers' shape. Biochemists, and in particular Roberto Strom, like to represent molecular equilibria and the dynamics of these changes in symmetrical equations [35]. In this case, the representation is particularly fitting. It shows that there are two series of events that can be brought about from the picture at the top-left to the bottom-right corner of the square in Figure 21. One depicts active induction of changes, the other selection/stabilisation of self-reached equilibria. However, the final result is the same.

The upper part of the scheme in Figure 21 depicts a mutant molecule: the horizontal arrows show that the equilibrium is slanted towards the inactive conformation (grey square), while the wild type active conformation (orange circle) represents a minority of the mutant. By adding antibodies specific for the wild type, two different changes may occur: i) the antibody binds at low level of affinity with the mutant conformation (left side); this situation is not satisfactory, and a low free energy ban is only reached if the wild type conformation is induced; ii) alternatively (right side), the antibody binds to the few

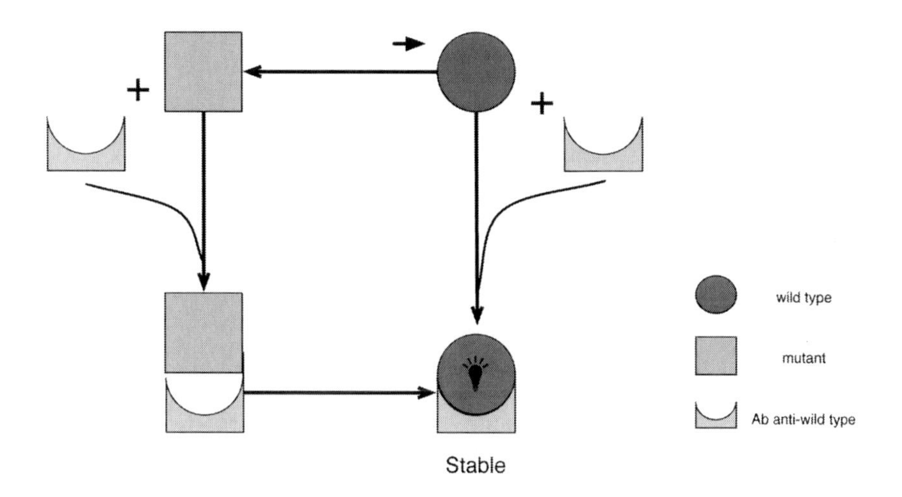

Figure 21. Equilibrium of a general antibody-mediated activation of a functional molecule. (see description in the text.)

Colour image of this figure appears in the colour plate section at the end of the book.

molecules still in wild type conformation. These fall out of the equilibrium altogether, and the latter is restored by a limited shift to the right that restores the 'red dot'. Newly formed wild type shapes are bound and stabilised by the antibody, and this time-consuming process continues. There is no way to predict if one of the two modes of activation is preferred, or whether both modes are enacted. Reaching the lowest free energy level, in any case, satisfies the energy equilibrium.

1.55 More Fields for Models, and Where to Find Them ♞

Of the Beta Gal mutants, several are switchable by antibodies. Another antibody directed to the wild type has an effect on the wild type tetramer itself: it stabilises the shape of the wild type and protects it from heat denaturation (the breaking point is raised by ten degrees Celsius). Activation and protection from heat are deeds of different antibodies among the anti wild type binders.

From this system, principles have been learned which are also valid for all conformational applications. Take an antibody whose paratope would fit perfectly to a wild type epitope. If this antibody meets a molecule whose conformation has been altered by mutation, chances are it will have no affinity at all for the remnants of the disturbed epitope. Another possibility is that it will still bind to the site, albeit with low affinity. This situation lacks stability, and represents a 'malaise' expressed by high level of free energy about the molecule. If there is the slightest possibility of a return to a wild type similar shape, and thus for the antibody-antigen bond to increase the affinity, a conformation change will happen that will restore all previous conditions including, in the present case, the catalytic function.

In other words: an antibody carries in its paratope the mirror image of an ideal epitope and craves to find it. If the ideal is corrupted, it will try its best to rebuild it. If it is in good shape, it will protect it.

Proteins that form or are part of receptors for cells or viruses are professionals of the conformational change: it is safe to assume that the conformation of the receptor is different before and after it has bound the counterpart that it recognises. It follows that it is possible to modulate the receptor function, if monoclonal antibodies are formed specific to the two conformations of the receptor. Anti-empty and anti-occupied Abs will block or will favour the receptors' binding.

By now everybody should be convinced that conformation is essential in immunology and therefore should be considered in the discrete models. There is certainly more than one way to do it, together with the upgrade of the description and simulation of nonlinear effects. Binary strings have served and do serve to simulate recognition and attraction between paratopes and epitopes; being linear, the strings accurately simulate *sequences* of amino acids forming an epitope. If it loses one out of four amino acids, such a linear epitope will probably drop some affinity. If it loses two out of four, it is bound to drop some more affinity. A similar loss suffered by a conformation-dependent epitope may possibly remain intact in its function or, if the *shape-shift* of the whole protein occurs, may well disappear from the universe of the epitopes altogether. It is clear that different measures must be used to record the two kinds of phenomena. For instance, a binary string can be designed with capabilities that can replace the old type, and use them to represent conformational effects and nonlinearity in the epitope. It is called *sliding string*. Before the *shift*, it faces the paratope's string, and the binding affinity is recorded. Then the shift occurs, in the form of the deletion of one bit; the loss is repaired by issuing the order 'all bits squeeze to the left'. The result of this can be seen in Figure 22.

Modelling conformation is a long stride for IMMSIM modellers, but it is also one of the most interesting next steps. Since protein conformation is involved in all biological processes, and at several levels of magnitude, one must be prepared to end up

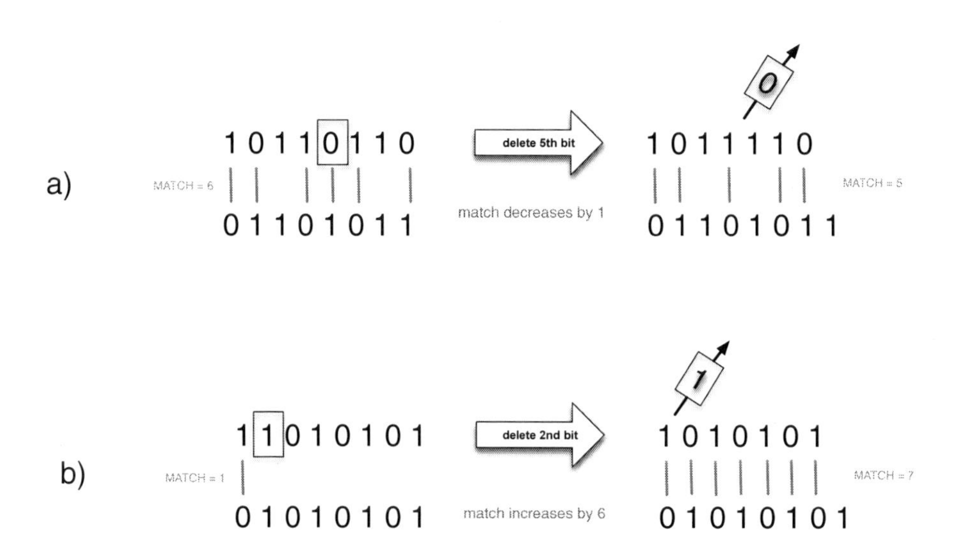

Figure 22. Binary strings can be used to identify objects and determine their mutual attraction (IMMSIM uses 24 bits for each paratope and each epitope, and for each receptor and each specific ligand). When these objects come in proximity, the respective strings align and each of their bits meets their opposite one. Attraction is only exerted between 1 and 0 and 0 and 1 (bit match). The sum of matches determines the affinity of the interaction. In the scheme, one vertical red bar symbolises a match.

Colour image of this figure appears in the colour plate section at the end of the book.

with many variations of the same model, or even with different models altogether. There are several pathologies that are now labeled prion diseases: Amyotrophic Lateral Sclerosis, Scrapie, Kuru. They all affect parts of the nervous system, and they all are characterised by infective conformation mechanisms, even if the pathogenic molecules are quite different in the various diseases. The modeller's problem in this early stage is not to mimic the various molecules according to their chemical nature. Instead it is important to prepare the discrete model to become able to represent the extreme *non-linearity* of the effects that a change of conformation can produce, compared with most 'immunological' contacts.

In the scheme of Figure 22 can be seen, on the left, two 8-bit-long pairs of strings. The total affinity is 6 for the upper combination (a) and 1 for the lower (b). If the strings are considered *solid*, any one-bit change in the string will decrease or increase the binding strength by just one unit.

To model *nonlinear* effects produced by shape shifts and represent, for instance, a prion that is hit by a point mutation and totally changes its attraction to the neighboring molecules, a new use of the binary logic is under consideration. Instead of *flipping* one bit, one can *delete* it, and *slide* the remaining part of the bit-string towards the left to fill the gap (elimination by mutation is observed during recombination of antibody variable parts and affects the paratope's affinity). The results of the two examples shown in Figure 22 are interesting; while the upper combination (a) suffers a decrease of one unit, the lower one (b) sees the affinity rise from 1 to 7. Since the effects of conformational changes are never *linear*, this new tool is the proper way to represent them.

1.56 A Note to Encourage Cross-Referencing

In the best of all possible worlds, science proceeds in its quest for knowledge utilising all well-known and easy tools. If biology struggles to reach the next level and is disturbed by the growth of complexity, then perhaps this world is not precisely that of Candide. One counter measure is to use mathematical tools to construct models of 'pieces of nature', then study them as if they were scientific hypotheses to be tested and eventually accepted or rejected. Since the model is simpler than nature, if the model results are accepted, a step in the direction of decreasing complexity is made. However, there are many ways models can be used in research, and IMMSIM has been applied to many problems over its lifetime. When one describes, as in this publication, the functions of the immune system, it is amusing to note how many distinct and different relationships have been realised between the model and any aspect of the immune system being described or discussed. The model may have mimicked an aspect; it may, in the process of this, have expressed a hypothesis about the relative importance of sub-functions; it may have systematically studied the effect of the dose of an antigen; it may have predicted the outcome described. And some of this chat may be interesting to some readers. Therefore black chess pieces are used as a simple code. The code list is reported and explained in Table 2.

Table 2. A chess code is used in the sections' titles to indicate whether the concepts discussed therein have inspired, have been incorporated in, are simulated by, are predicted by, or are used as validations for IMMSIM.

	Subsections	Chess code
Inspires IMMSIM	1.3 1.6 1.9 1.10 1.11 1.14 1.17 1.18 1.20 1.21 1.22 1.26 1.27 1.30 1.31 1.32 1.33 1.35 1.36 1.37 1.38 1.40 1.41 1.42 1.43 1.46 1.47 1.52 1.53 1.54 1.55	♞
Incorporated in IMMSIM	1.7 1.8 1.9 1.10 1.12 1.17 1.20 1.21 1.22 1.23 1.28 1.30 1.31 1.32 1.33 1.35 1.37 1.39 1.43 1.44 1.46 1.48	♜
Simulated by IMMSIM	1.11 1.14 1.18 1.23 1.24 1.26 1.30 1.35 1.37 1.41 1.42 1.46 1.48	♝
Predicted by IMMSIM	1.19 1.20 1.22 1.23 1.31 1.41 1.49 1.50	♛
Validates IMMSIM	1.19 1.31 1.49 1.51	♚

1.57 Modelling to Learn Immunology

Although the principal role of modeling the Immune System is to advance Theoretical Research by producing predictions and hypotheses, its potential educational role in Immunology is well established. In 1992, Celada and Seiden wrote a book chapter titled "Teaching Immunology: A Montessori approach using a computer model of the Immune System" [36].

Computational models, like IMMSIM, proved to be valuable tools in two official Immunology courses over the years 1996-98: one at the University of Genoa, taught by Franco Celada (with Roberto Puzone, Claudia Calcagno and Dario Ghersi), and the other at Princeton University, taught by Martin Weigert (with Philip Seiden, Steven Kleinstein and Jeffrey Stewart).

The participant students downloaded a modifiable version of IMMSIM, and their assignment was to simulate one function of the immune response. They had to study consensus, upgrade the code, run the simulation in their laptops, compare results of alternative paths, and select by trial-and-error the most efficient setup. At the end, individual solutions were run in parallel, discussed and evaluated. Why does this method have an edge over conventional teaching? Tentatively, on two counts: ***a)*** according to Giovanni Battista Vico (b. 1668), complete knowledge is reached only of what one has made himself (an hyperbole with a grain of truth); and ***b)*** modern scientific methodologies produce an increasing mass of data, whose specific importance is increasingly difficult to evaluate: simplification, perspective and selection by performance are all required in modeling, and by thwarting man-added complexity can make both Biology and Models easier to grasp.

A different educational event, in the form of a seminar spanning over one month and focusing on interdisciplinary encounters, took place in 2001 at the ZiF Institute of Bielefeld University in Germany. Among the Modellers were Filippo Castiglione, Brynja Kohler, Roberto Puzone, Rob De Boer,

Michele Bezzi, Stefano Ruffo, and among the Immunologists, N. Avrion Mitchison, Eli Sercarz, Franco Celada, Claudio Franceschi.

1.58 An Unconventional Glossary for Aliens

Items in the glossary are referred to by the sections where they appear in the text.

1.2

Phagocytosis is the ingestion (and destruction) of foreign particles, other cells, and bacteria by specialised cells. In the immune system, the phagocytes are macrophages and other antigen-presenting cells. They process the internalised object by enzymatic digestion and may present the resulting fragments on special receptors on their membrane.

Specific memory is the property of the vertebrates' immune system to produce a different response to an antigen if it is presented again, some time after the first encounter. The response is faster, stronger and, in case of infectors, can prevent their developing the infection altogether. Memory is affected by a population of ready-to-respond cells which stem out from the growing clones during the primary response and are then stored in the lymphoid organs until a new antigenic stimulus reaches them.

1.6

Tolerance: Specific tolerance is the failure to—or the decision not to—respond to an antigen. Tolerance to self-antigens, acquired in early ontogeny, is an essential feature of the immune system. Tolerance induced in adult life, by exposure to antigen excess or chemical immunosuppression, as in organ transplantation, may never be complete.

Chimerism: In Greek mythology, the Chimera is a monster with three heads (a flaming lion's in front, a goat's on the back, and a serpent's at the tip of the tail). Chimerism in biology means presence of genetically diverse cells or tissues in a single organism.

1.7

Lymphokine, lymphokine receptor: Lymphokines are cytokines (small proteins that carry functional signals to cells engaged in defence related deeds) produced by lymphocytes. They act by interacting specifically with receptors on the cell membrane. Correlated signals induce the secretion of the Lymphokine and the appearance of the receptor. These mechanisms are used in most signalling and control of immune responses.

MHC (major histocompatibility complex): A large cluster of genes coding for most peptides of histocompatibility antigens, they emerged in a dramatic series of findings in the 1960s, to function as antigen receptors for immunocytes of humoral and of cellular responses. The MHC cluster is located in human chromosome 6 (in mice, chromosome 17) and is transmitted as a unit, with rare recombination. This unit is called *haplotype* and the typing of HLA (human leucocyte antigen) specificity, essential for the donor-recipient matching for transplantation, is achieved when both haplotypes of donor and recipient are determined.

TCR (T-cell receptor): This is the membrane-bound receptor of all T-cells. It consists of two chains, alpha and beta, each formed of two hundred amino acid immunoglobulin units. Each chain's outer unit is variable, and the noncovalent interaction of the alpha and beta variable parts forms a paratope that will bind the antigen's peptide if it is presented by an MHC recognised by the TCR. Thus the bond is called MHC-restricted.

1.8

Paratope, epitope: In antibody, or B-cell antigen receptors, the paratope is the three-dimensional space delimited by the interacting variable parts of the heavy and light chains. The epitope is the region of the antigen that interacts with the paratope and consists of about four amino acids or about six sugar units. If the amino acids are sequential units of the same protein string, the epitope is linear; if the amino acids belong to two or more different protein strings or different regions of the same string, the epitope is conformation-dependent, or *conformational.*

1.9

APC: Antigen-presenting cell. APC is any cell, lymphoid (B), dendritic or macrophage that, having ingested and processed antigens, is capable of presenting specific peptides on MHC receptors to Th or T effector cells. Antigen presentation is a key function in both humoral and cellular responses.

1.10

Syngeneic: Syngeneic means genetically identical as are mammalian uniovular twins and rodents produced by directed inbreeding. The latter, after 20 cycles of brother-sister breeding, have a 99.9% probability of genetic identity and therefore accept transplants from members of the same strain. In addition, inbred mice are homozygous for all loci.

1.14

Hydrophobic, polar attractions: These are the 'weak' forces that allow all noncovalent interactions between molecules and notably between proteins, where the same forces (as attractions or repulsions), when exercised between different atom clusters of the same or adjacent molecules, determine their conformation in space. Polar forces are sparked between electrically charged regions, as +/– attractions or +/+ and –/– repulsions, while hydrophobic attractions are generated between two nonpolar groups trying to extrude water molecules. The strength of weak forces can increase by several orders of magnitude when many active groups near each other participate, as happens in the immune system for receptor-epitope interactions.

1.15

Darwinian Evolution: The hypothesis of natural evolution put forward by Charles Darwin in the mid 19th century, based on the observation of continuous generation of diversity and selection for fitness in the environment, resulting in the evolution of animal species. Immunology confirms the Darwinian principles in all phenomena of somatic evolution.

1.16

Cell cooperation: Any functional interaction between individual cells of different lines in an organism, typical within the immune system. The label has been first used to describe the Th–B interaction allowing the completion of the humoral response.

Immunogen: An antigen that, besides being able to bind specific antibodies via *its epitope(s),* is *also* capable of eliciting a primary immune response. For this function, the antigen must possess a number of *peptides* to be presented to helper T-cells' receptors, and be *nonself* with respect to the responder. By a 'rule of thumb', foreign proteins of molecular weight greater than 10,000 Daltons are immunogenic.

Cytokine storm: In the process of transmitting the immune activation, many different cytokines—secreted by different cells—are required, and interact.

Covalent/noncovalent bonds: In proteins, strong atomic interactions known as **covalent bonds** involve exchange of orbital electrons, such as the interaction -COOH + H_3N- which results in the *peptide bond* -OC-NH-, the key of the protein chains, making both *the primary and secondary structure.* **Noncovalent bonds** underpin the *tertiary* (three-dimensional) structure and the *quaternary* interactions between distinct molecules or between *monomers* in *polymeric* proteins; all these are manned by the so-called *weak attractions* (a variety of *electrical* and *magnetic* attractions between molecular groups, and the *hydrophobic* forces that appear between water-repelling groups. The *noncovalent* forces, thanks to their flexibility and ubiquity allow most essential biological functions, from enzymatic cleavage of substrate to immune interactions between antigen and antibody.

Inter- vs intra-molecular help: If the connection between the *helper T (Th)* and the *helped* (B or DC) cells uses elements *within the same molecule*, we are talking of *intra-molecular help*. Instead, in case the elements belong to two different molecules, kept in functional proximity by *noncovalent bonds*, we are witnessing an *inter-molecular help (IMH)*. This mechanism has inescapable effects. For example, when the *epitope* is on the Fc of an antibody, and the *foreign peptide* is on the antigen that the antibody binds, nature's carefully engineered test for *self* can be defeated and an autoimmune response can break out.

Rheumatoid factor: This is an antibody formed against circulating *self* antibodies. It is practically ubiquitous, but it is found in high and *diagnostic* titers only in lupus and a variety of rheumatic patients. Rheumatoid factor (RF) is a typical *inescapable* autoimmune response made possible by spontaneous *inter-molecular help*.

CD4-gp120: CD4 is a receptor and an eponymous marker for the helper T cell family (Th1, Th2). gp120 is a surface protein of HIV, the retrovirus responsible for AIDS. There is a strong noncovalent attraction between these two proteins, making CD4 the site of entrance for HIV into a human cell. Soluble CD4-gp120 complexes are found in the serum of AIDS patients. A change of conformation of the complementarity-determining region (CDR) occurring at the moment of binding gp120 has been described [3].

1.17

Apoptosis: Programmed cell death; in lymphocytes it is enacted in a number of circumstances 'for the good of the immune system', by activating death and DNA degradation process.

1.18

Chemotaxis: The property of certain cells or primitive organisms moving in body fluids to be attracted or repelled by chemical substances.

1.22

Two-signal hypothesis: In the mid 1960s, this hypothesis replaced the idea that merely binding the specific antigen stimulates specific immune responses. The second stimulus, better called co-stimulatory signal, is a requirement for the successful activation of T-cells which recognise via their TCR specific antigens presented by the MHC of dendritic cells. Co-stimulatory instruments are CD4 (in case of Th) and CD8 (in case of TC). On the other bank, the dendritic cell produces stimulatory molecules whenever it receives a 'damage' signal or perceives the presence of microbial components. A successful co-stimulation (also required in the cell cooperation Th-B) testifies to the seriousness of the infection and lifts the block normally in place for fear of producing damage to the body by responding. The co-stimulation combines the two essential reasons for the body to start the defence: the assurance that the antigen is not self, and that it represents an aggressive agent. Umberto Eco has considered the negotiation between different actions/decisions before launching the response as the sign of a cognitive system.

1.25

Variable and hyper variable regions: The long immunoglobulin units of all 116 amino acids are coded by copies of the same gene and are therefore identical (*constant*) in all antibodies of an individual, with the exception of the N-terminal units of both the heavy and the light chain that are *variable*. Variable means that a large proportion of sequence positions have different amino acids' occupancy in antibodies from different B-cells or clones. The Variability index (number of different AA in the same site found in hundred observations) peaks in three clusters in both heavy and light chains, around position number 32, 53 and 95 of each variable region (the clusters are called *hyper variable* regions, or, since they sport the amino acid that will eventually make contact with antigen, CDR: complementarity-determining regions 1, 2, 3).

1.26

Idiotype: Discovered in the early 1960s by Jacques Oudin and independently by Henry Kunkel, the idiotype is the 'outer aspect of a paratope', and is therefore an epitope correlated to the paratope's specificity. One paratope expresses several epitopes (idiotopes); the ensemble of the idiotopes pertaining to one antibody is its idiotype. When antibodies are raised against an antibody (Ab1) some (Ab2-beta) are called 'internal image of the epitope' because they share the original antigen's specificity. It has been proposed that Ab2-beta be used in vaccination. The majority of Anti–Ab1 do not bear similarity to the antigen, and are labelled Ab2-alpha.

Lupus: Systemic lupus erythematosus is a disease related to the rheumatoid family with distinct capacity to produce many autoimmune antibodies. One of them, anti-DNA, cross-reacts with DNA of all species, and its presence is used as a diagnostic criterion.

Lethal irradiation is dose of any radiation, the exposure to which causes death. To measure radiation lethality, groups of rodents are exposed to increasing doses until the dose causing the death of 50% of the animals in ten days is determined. An alternative method, using only one mouse and determining when it was 'half dead', as mentioned by an English researcher during a meeting in Oak Ridge, Tennessee, in 1959, was considered a rowdy joke, and it was.

1.27

Dendritic cell: Distinct strain of these cells, generated in the bone marrow and belonging to the extended family of macrophages, present antigen to B-cells in the lymph node follicles and to immature lymphocytes in the thymus, and manage the entire activation process of all immunocytes.

Brigitte Bardot: The famous French movie star of the 1950s–1960s retired to her villa in St Tropez in 1973. There, animal welfare became her prominent mission and remains so to the present day. What makes her happy to appear in this book is the hope—shared by us—that the growing use of models in Biology will reduce the need of animal testing.

1.29

Agglutinin: Any antiserum or antibody that causes *agglutination* when added to a suspension of erythrocytes. Agglutination is a simple diagnostic test in blood typing, but can be used by binding unrelated antigens to blood cells on the cell membrane and using antibodies to those antigens as agglutinins.

Lymph vessel: The network of vessels, similar but independent to the blood vessels, that allows the circulation of lymph and carries all immunocytes, macrophages, dendritic cells and lymphocytes from the peripheral tissues to the lymphoid organs and vice versa. The heart indirectly provides the circulation within the lymph vessels, since the final lymph vessel joins the thoracic vein.

1.31

Vaccine: Antigen-related substance administered to provoke a preventive immune response and/or build up memory that may be triggered by an eventual infection, with the purpose of causing a rapid disposal of the pathogen. Since pathogens may be dangerous, the vaccine may be a cross-reacting molecule, or a dead or attenuated virus or bacterium. The word 'vaccine' (*Vacca* is Latin for cow) refers to the use of cow pox (which is not lethal for humans) to protect potential patients from smallpox in the successful 'first' by Dr Jenner, in England.

1.33

Navy Seals: Surname of leatherhead corps of the U.S. Navy that became popular in 2012 for their daredevil commando actions.

1.35

Immunoglobulin class: A family of glycoprotein synthesised by the lymphoid cells which makes up all structures used within the immune system as structures able to recognise antigens, exchange information and manage immune responses. They serve as carriers of specific strings of enormous diversity without disturbing the behaviour, elasticity and robustness of the individual units.

Class-1, class-2 receptors: The MHC molecules comprise these distinct types of receptors destined to present antigen peptides to the TCR of i) T cytotoxic cells and ii) T helper cells. They are similar in size and appearance. Class-1 receptors consist of a 3-unit membrane-bound chain containing two variable parts, and of a 1-unit chain that is not membrane-bound. Class-2 consists of two 2-unit chains, both bound to the membrane and carrying one variable part.

HLA: The human leucocyte antigen system is the human MHC. Its mouse counterpart is the H2 system.

Cognitive system: A cognitive system is any combination of entities that originate, receive, and interpret a signal or sign, capture its meaning and determine a proper reaction. Cognitive systems are the application field of semiotics. Semiotics is the discipline of these studies. In biology at large, simple reflexes (hit the knee—obtain the kick) are outside semiotics, while for example, the cell cooperation between B and Th cells that ends a period of hesitation of the former when the latter proves that its attack is not anti-self, is considered the epitome of a cognitive system, an unconscious brain.

1.40

Cytokine: See under Lymphokine.

1.43

Tinkerer: Together with the French 'bricoleur', this word has been used to characterise the shifty solution often adopted by evolution (Francois Jacob), and fits especially well in describing the upgrading of the immune defences to the specific immune system.

Complement: A number of glyco- and lipo-proteins present in the blood serum, many endowed with enzymatic properties and able to mount sequential reactions that can end in the destruction of cell membrane, causing killing of bacteria and haemolysis of erythrocytes when recognised by antibodies. This potentially lifesaving, non-specific defence is triggered by precise anchors, such as the C1q receptor on heavy chain of the antibody. The receptor pulls the trigger when the antibody has both paratopes specifically occupied. The complement is a non-specific bomb guided by a specific aiming device.

1.44

Ribosome: Cellular organ hosting the synthesis of all proteins by sequential binding of amino acids, according to the code inscribed in the genome.

1.46

Phylogenesis, ontogenesis: Both terms indicate the developmental history. Phylogeny is concerned with the species, while ontogenesis is limited to the growing and aging of an individual.

Catch-22: In Joseph Heller's book with this title, a war pilot who flew too many missions claims insanity to get off the hook. But a 'Catch-22' of the military law considers his claim a highly rational act, and on this basis rejects it and orders the continuation of missions. A contradiction of logic that cannot be resolved. (Another one is the sentence "I am a liar". Is it true or false?)

1.51

Antigenic sin: An immune response triggered by an antigen that happens to cross-react with a previous immunisation or vaccination, even many years in the past. The response produces antibodies or cells specific to the primer, instead of the challenging antigen. Antigenic sin is one expression of malfunction of the immune response caused by overpowering memory. It results in low-affinity and low-specificity responses.

Immunopathology: For the lack of a more precise word, immunopathology designates the state of chronic inflammation that follows the termination of a viral infection, and can cause all sorts of additional damage to the target organ and tissues. Autoimmune activation has been shown, and it has been suspected that the chronic inflammation is favoured by low-affinity anti-viral responses and by the precipitation of low-affinity immune complexes. This makes immunopathology a symptom loosely related to immuno-senescence, the aging of the immune system.

1.52

Lactose operon, allostery: Lactose operon is a group of adjacent genes on the chromosome of *Escherichia coli* whose functions are correlated to the digestion of lactose. These are gene Z, beta galactosidase, gene Y, permease, and gene A acetylase. Their regulation utilises a molecule, the repressor, which, when sitting on the chromosome, blocks the promoter, and thereby the enzyme synthesis. But the suppressor is susceptible to a conformational shift that can be induced by a small lactose-related molecule; thus, it loses grip and allows the synthesis. This mode of signalling the presence of substrate and initiating the production of the three enzymes has been called *allostery* by Jacques Monod, and is the first demonstrated model of gene regulation.

1.54

Porsche: The German cars, built from 1948 by Eng. Ferdinand Porsche, have always been examples of high speed and quality, and status symbols. Citing Janis Joplin: "Oh Lord won't you buy me a Mercedes Benz?—My friends all drive Porsches—I must make amends!"

1.55

Prions are a recent discovery, somewhat astonishing vectors of disease. They are involved in Kuru (the slow infective degeneration of nervous tissue), in Creutzfeld-Jakob disease and in the 'mad cow' disease. There is a waiting list of other candidates. Prion is purely protein, and may assume alternative conformations, non-functional or 'toxic'. The latter conformation is in contact with other, normally conformed molecules; it induces in them the alternative, non-functional conformation, initiating a domino effect. Non-functional molecules are deposited in large amounts and severely disturb the nervous tissue function.

2

Aliens for Immunology

The book now departs from the viewpoint and language of immunology and adopts those of a computational modeller. The writer is an *alien* who happens to meet an immunologist at a time when computational immunology was unsuspected in Italy. He will first tell how this meeting took place. Then, in the following sections, he will show how the immunological concepts described in the preceding sections have been translated into a computational model. Successively, he will illustrate the model at work, that is, simulations' results will be commented on to give a sense of how one can experiment with it. The choice of simple examples is not casual but is aimed at keeping complexity at bay. Alien modellers know that the usefulness of a mathematical model is to be found halfway between triviality and extreme complexity. To give the impression that IMMSIM is exceedingly complicated would be a fault. The correct key to understanding it is to view it as the *holistic* combination of simple sub-models, each enacting a 'concept' or describing a 'phenomenon' of the immune system. These concepts are nothing less and nothing more than those reported in the first part of this book. It is now time to see them in action.

2.1 The Unusual Mix of Immunology and Computer Science

Mathematical Biology has a long tradition dating back to Alfred J. Lotka in 1910, and Vito Volterra in 1926, who independently developed the first model of the predator-prey relationship. Apart from modelling ecosystems, not much had been developed for a long time, until scientists of 'exact' disciplines like physics and mathematics turned their interest to biology in general and immunology in particular. That was the time when Franco Celada decided to embark on this interdisciplinary endeavour and got in contact with Phil Seiden (of interest to note, the initiative came from the immunologist and not from the physicist). This story has been told in the foreword.

A few years later, Filippo Castiglione stumbled on this interdisciplinary adventure. He went to his professor Gianpiero Cattaneo at the Computer Science Department of the University of Milan, and asked for an interesting topic for his master's thesis. Cattaneo, who had recently been spurred by Franco Celada to find a student who was willing to work on IMMSIM, asked, "Do you want to go to New York?" What a question!

At first Filippo thought that, as a computer scientist, he could not have chosen a more remote assignment than modelling and simulating the immune system. At that time he was inspired by topics like artificial intelligence, connectionist models, cellular automata and complex systems. Biology was out of sight. However, as soon as he started looking into it, he realised that IMMSIM was a cellular automaton and the immune system, indeed, a complex system. Encouraged by these facts, he took the challenge and embarked on this adventure together with biologists and immunologists. Everything sounded very unfamiliar to him. He was supposed to talk about Valiant's theory of learnable and Lyapunov functions whereas he, most often, was hearing words like lymphocyte, epitope, germinal centre and Burnet's clonal selection theory. What was that?

Luckily the alien Phil Seiden was around. He was talking the same language! Although Filippo's English was rather inadequate, he was able to read the equations Phil wrote on a piece of paper the day he met him for the first time in Milan. That was inspiring. In less than half an hour of discussion he not only figured that these would be the topics of tomorrow but also decided to use IMMSIM to model hypermutation and the immune response against a mutating antigen such as HIV (see Section 1.19). Everything that followed made up his thesis.

Soon enough, the difficulties arising from the use of the APL2 language suggested the idea of recording the whole IMMSIM program in another computer language. So, after few months, he came up with a beta version of C-ImmSim written completely in ANSI C and following the what-would-be-called agent-based paradigm.

While IMMSIM had been conceived as a stochastic cellular automaton, C-ImmSim was coded keeping in mind ideas like Marvin Minsky's frameworks [37, 38] used in artificial intelligence and object oriented programming (although the computer language adopted was not C++). Other guiding principles were Christopher Langton's artificial life [39] and John Holland's Genetic Algorithms [40, 41]. He imagined lymphocytes as complex agents interacting in a certain environment and evolving from generation to generation, following a kind of 'survival of the fittest' principle. Eventually, he got concepts from artificial intelligence and optimisation mixed up in an artificial life computer program of a very complex system. He couldn't have asked for anything more appropriate. This was exciting!

At once, the C code became extremely complex and demanding in terms of resources. But it was flexible, easily extensible and general. It was a model of immunological models. The original aim of Franco and Phil, to create an *in silico* tool to explore immunological theories was—and still is—their innovative gift.

2.2 Classical Modeling Techniques Versus the 'New Kind of Science'

The great majority of models of the immune system are based on ordinary or partial differential equations, with the number of the former much greater than that of the latter. The value of the first mathematical models of the immune system is purely academic while more recent models have been shown to have substantial predictive ability. An exhaustive review of the early development of mathematical models in immunology is beyond the scope of this book and can be found elsewhere [42-46].

One of the first mathematical (equation-based) models in the domain of immunology was that of Hege and Cole in 1966 [47]. It described the change of circulating antibodies depending on the plasma cells. In early 70s, Jilek [48] proposed different probabilistic (Monte Carlo[4]) models of B-Ag interaction with clone formation. G. Bell published many articles by using mathematical models of Burnet's clonal selection theory [9] to model the humoral response to multivalent antigens [49]. He further expanded the model by allowing the antigen to multiply, where the antigen-antibody interaction was described in terms of a 'predator-prey' system [50] (the first predator-prey model is the well known Lotka-Volterra model [51, 52] mentioned above). In 1975, Bruni et al. [53] suggested a model of humoral immune reaction, taking into account the concept of heterogeneity, hence the affinity, of immunocompetent cells. In 1978, Asachenkov [54] suggested a model of the cooperation between B and T and the production of two classes of immunoglobulins. In 1975, Richter [55] and Hoffmann [56] were the first to propose models based on the idiotype network of Jerne [10, 11]. Another interesting step in the development of new mathematical models was made by Waltman [57], who adopted the concept of states and state switching in the representation of lymphocytes. In 1977, De Lisi [58] published several papers on the mechanisms

[4] The Monte Carlo is a numerical method that uses random numbers to perform a simulation. It has been used for centuries, but only in the past decades has the technique gained the status of a full-fledged numerical method capable of addressing complex applications. The name 'Monte Carlo' was coined by Metropolis (inspired by Ulam's interest in poker) during the Manhattan Project of World War II, because of the similarity of statistical simulation to games of chance, and because the capital of Monaco was a center for gambling and similar pursuits.

of tumour-immune interaction starting from Bell's simple model. In 1976, Dibrov et al. [59] took simple models but added the concept of space to analyse the effects of diffusion and time lags on the immune-pathogen dynamics.

A series of models used the shape-space approach. Each point of the multi-dimensional Euclidean space is associated with a different receptor's generalised shape and each coordinate represents one of the main aspects involved in pattern recognition. Stewart and Varela were the first to propose such models [60]. They used a bit-string model in a 2-D shape space to analyse the meta-dynamics in which new clones are constantly generated in the bone marrow.

All of the above mentioned models represent the changes of a certain quantity, like the number of cells or the concentration of a molecule in a compartment of the immune system, as a function of the birth and death rates and of other kinetic parameters used to describe their interactions. For example, the so-called AB-model [61] used in theoretical studies of Jerne's immune network (see section 1.26) describes the dynamics of B-lymphocytes ($B_i(t)$) and antibodies ($A_i(t)$) for each i-th clone ($i = 1 \ldots M$):

$$\frac{dB_i}{dt} = m + B_i(pf(h_i) - d_B)$$

$$\frac{dA_i}{dt} = sB_if(h_i) - d_ch_iA_i - d_AA_i$$

$$h_i = \sum_{j=1}^{M} J_{ij}A_i$$

where m is the rate of maturation of B-cells by the bone marrow; p is the division rate of activated lymphocytes; d characterises the death rate; s is the secretion rate of antibodies; d_c is the elimination rate of antibody-antibody complexes. Idiotypic interactions are mediated through a field h_i, which is determined by the concentration of antibodies A_j and by the affinity J_{ij}. The activation function $f(h_i)$ is an important ingredient of the model and it determines the intensity of the stimulation of the clone i

by the idiotypic field h_i. These equations are used to study the dynamical regimes: oscillatory, chaotic, etc.

Modelling natural phenomena by means of differential equations has always been the *way of doing*. However, in the use of differential equations to model the immune system, one can spot three main limitations that are worth a moment of reflection. First, differential equations assume the existence of sufficiently large populations of homogeneous entities. Yet, each cell of the immune system has a unique life history that defines its interaction with the environment, so that averaging over populations is often inadequate. The typical solution to this problem is to divide the cells into classes according to a small number of characteristics, but this approach ignores the true complexity and a myriad of special cases so important to experimental immunologists. The second limitation is that equations describe only the average behaviour of a system. Although there are questions for which knowing the distribution of behaviours is not relevant, there are many more questions that cannot be addressed without this knowledge. Finally, if models want to capture the quintessential complex nature of the immune system, they must necessarily involve nonlinearities; this makes the analytical solution of differential equations difficult [62].

On the other hand, the advantage of adopting a 'discrete' modelling paradigm such as the agent-based modelling (see next paragraph) lies in the fact that the behaviour of the cellular entities originates from fundamental occurrences in biological processes—for instance, the binding of molecules to receptors—and as such, they are more intuitive and easier to implement as computer instructions. Moreover, one is able to represent components and processes of interest in biological language so that the approximations in the model are more biological in character than mathematical. Another advantage of this paradigm is that, since nonlinearities are not intrinsically difficult to handle, it is easy to modify the complexity of the interactions without introducing any new difficulties in solving the model. Last but not least, a discrete model naturally allows for stochasticity, since the events regulating the dynamics are designed as probabilistic rules, giving rise to highly random agents' behaviours [63].

2.3 From Spin-like to Agent-based Models

C-ImmSim belongs to the class of 'discrete' models. It can be described in different ways. It can be seen as an Agent-Based Model (ABM), but also as a generalised spin model (also called *Ising model* from Ernst Ising [64]), a Lattice Gas or a Cellular Automaton (CA). All these paradigms share the same characteristic of representing the 'particles' or agents individually. Indeed, the agent of the ABMs itself can be considered an extension of the concept of *spin*. It was the widespread use of powerful computers that allowed for the addition of more realistic descriptions of the internal state-space of the spins (namely, the entities or agents) and of complex interaction rules. In ABMs, the procedures are much more than simple Boolean rules as in the CA-type model from Kaufman [65] (see below); they are sophisticated algorithms. Interestingly, while physicists tend to refer to spins and cellular automata that use minimalistic Boolean rules to model disparate phenomena (immune response included), computer scientists follow the *artificial life* (ALife) paradigm as a source of inspiration for the agent-based modelling [66], and the rules they come up with to describe what the immune cells do cover more biological details. This is greatly appreciated by the immunologists. However, in practice, the differences between spins and artificial life's agents are apparent only because ALife largely adopts concepts of complex system simulation such as spins and automata.

This 'modern' approach to the simulation of complex phenomena lacks an underlying mathematical theory and, to date, a formal structure. While simpler models like Ising or cellular automata have been studied not just numerically (i.e., running them on a computer) but also analytically, and general conclusions in some manageable cases have been reached [21, 67], there is no *simple* and general mathematical description of agent-based models [68]. Likewise, to date, an attempt to formalise the Celada-Seiden model (CS-model) is still lacking.

On a very basic level, every simulator can be described as a discrete dynamical system, where the deterministic or stochastic function $F : \mathbb{R}^n \mapsto \mathbb{R}^m$ represents the operations of the algorithm to update the state of the system $s(t)$ at time t, that is, $s(t+1) = F(s(t))$. Since the expressive power of the computer algorithms is higher than that of classical operators and functions, it is hard to express the *semantic* of a simulator by means of the latter. C-ImmSim is no exception. As a matter of fact, it is just the contrary. The complexity of the topic and the variety of processes, together with the heterogeneity of the entities involved, make the specification of the function $F(\cdot)$ a gargantuan and sterile exercise. The question is: would it help in decrypting the complexity of the immune system?

2.4 The Ancestors

It is useful, at this point, to walk through different historical models and get an idea of how these algorithms have evolved from simple and stylised representations to very complex and detailed descriptions of phenomena.

Many population models of the immune systems have been developed in the past [44, 45, 69]. All models start from either one of the two existing biological theories of the immune system, namely the *clonal selection* and the *idiotype network*. While continuous models have been formulated in the framework of both immunological theories [44-46], early discrete models were mostly based on Jerne's theory [11].

The main task of the immune system is to perform pattern recognition between cell receptors and antigens. The binding mechanism is based on different physical effects (short range noncovalent interactions, hydrogen binding, van der Waals interactions, etc. [46]). In order for a receptor and the molecule that it binds, a ligand, to stick together, there must be extensive regions of complementarity. Perelson and Oster (1979) [70] called the constellation of features important in determining the binding among molecules the *generalised shape* of the molecules. Describing this shape as a point in a multi-dimensional space (the *shape space*) specifies the generalised conformation of a binding region. From this consideration, Oster and Perelson estimated that in order to be complete, the receptor repertoire should satisfy the following conditions: (i) each receptor must recognise a set of related epitopes, each of which differs slightly in shape; (ii) the repertoire size is of the order of 10^6 or larger; (iii) at least one subset of the repertoire size is distributed randomly throughout the shape space [46]. Later, Farmer, Packard and Perelson (1986) [71] introduced the use of binary string to represent the shape of receptors. To determine the degree of complementarity between strings, many string-matching algorithms are available. This representation has been used in several discrete models.

Discrete models of the immune system have been constructed using different techniques. Models based on *stochastic automata, cellular automata* and *lattice gas* go back to the last two decades. These models are used nowadays to describe biological phenomena and produce interesting results [72, 73]. The approach of a group in Santa Fe Institute and University of New Mexico (Stephanie Forrest and co-workers starting in early 1990s [74, 75]) brings the experience of computer scientists into the field. The guiding principle of this approach (which is not further discussed) is a deeper comprehension of the immune system in order to use its *information-processing features* in scientific applications (see the interesting review by S. Forrest and S.A. Hofmeyr [76]). This area evolved in the Artificial Immune System paradigm, devoted to finding practical solutions to applied problems requiring pattern recognition or optimisation methods (more details can be found in [77]). These algorithms get inspiration from the clonal selection theory, which is a Darwinian selection process. Therefore, they are not too dissimilar to the genetic algorithms.

Perhaps the model of Kaufman, Urbain and Thomas (KUT model) is the very first attempt to model the immune response by means of discrete mathematics. The original KUT model that was introduced in 1985 [65] considers five variables to represent immune cells and molecules: antibodies (Ab), helper T cells (Th), suppressor T cells (Ts), B lymphocytes (B) and antigens or virus (Ag). A Boolean variable denoting 'spin up' for high concentration and 'spin down' for low concentration designates each entity. The rules modelling the dynamic evolution of these variables are expressed by logical operations. The application of the rules is iterated over discrete time and the dynamics observed. The discrete evolution rules are:

$$Ab(t+1) = Ag(t) \text{ AND } Th(t)$$

$$Th(t+1) = Th(t) \text{ OR } Ag(t) \text{ AND NOT } Ts(t)$$

$$Ts(t+1) = Th(t) \text{ OR } Ts(t)$$

$$B(t+1) = Th(t) \text{ AND } (Ag(t) \text{ OR } B(t))$$

$$Ag(t+1) = Ag(t) \text{ AND NOT } Th(t)$$

where *AND*, *OR* and *NOT* are the usual logical (binary) operators. Although simple, this model shows a coherent dynamics. Of the $2^5 = 32$ points in the state space, the dynamics admits five fixed configurations. These identify the immune system global state: naïve, primed, immune, paralyzed, and sick.

Many others followed the KUT model. For example, Weisbuch and Atlan [78] focused on the special case of autoimmune diseases like multiple sclerosis, in which the immune system attacks the cells of the nervous system of the host.

As in the model of Kaufman et al., this model uses five binary variables representing: killer cells (S_1), activated killers (S_2), suppressor cells (S_3), helpers (S_4) and suppressors produced by the helpers (S_5).The different types of cells influence each other with a strength that is 1, 0 or –1. At the next time step, the concentration of one cell is unity if the sum of the interactions with the various cell types is positive; for zero or negative sums, the concentration is taken as zero, as seen in the formulas

$$S_1(t+1) = \mathrm{sgn}\left(\sum S_1(t) + S_4(t) + S_3(t)\right)$$
$$S_2(t+1) = \mathrm{sgn}\left(\sum S_1(t) + S_4(t) - S_3(t) - S_5(t)\right)$$
$$S_3(t+1) = \mathrm{sgn}\left(\sum S_1(t)\right)$$
$$S_4(t+1) = \mathrm{sgn}\left(\sum S_1(t)\right)$$
$$S_5(t+1) = \mathrm{sgn}\left(\sum S_4(t)\right)$$

where $S_i(t)$ denotes the concentration of the ith component at time t and the function sgn(x) defined on the natural numbers $\mathbb{N}$ is 1 if $x > 0$ and 0 otherwise. This model shows the existence of only two basins of attractions over $2^5 = 32$ possible states: the empty state where all the concentrations are zero, and a state where only activated killers disappear while the other four concentrations reach unity.

Further generalisations of this model consider the same dynamics but apply statistical physics, putting the cells on a lattice to allow simulations (Ising-like models). In Dayan et al. [79] the authors put five variables on each lattice site, corresponding to five Boolean concentrations (0 or 1). Note that this model is an Integer Lattice Gas with $r = 5$ (five entities) and $K = 2$ (two states per entity). Each site influences itself and its nearest neighbours in the same way as in the model of Weisbuch et al. For a square lattice of $L \times L$ sites, there are $r \times L^2$ spins. The main difference is that, in this model, the summation in the model equations runs over the site itself and its nearest neighbours. Interestingly, this lattice-version of the Weisbuch-Atlan model is found to have a simpler dynamics than the original model as the number of fixed points is found to be smaller than in [78].

Pandey and Stauffer further extended the KUT model using a probabilistic generalisation of deterministic cellular automata. Their model focuses on a possible explanation of the time delay between HIV infection and the establishment of AIDS [80, 81]. They represented helper cells (H), cytoxic cells (S), virus (V) and interleukin (I). The interleukin molecules produced by helper cells induce the suppressor cells to kill the virus. The dynamics is the result of the following rules:

$$V(t+1) = H(t) \text{ AND NOT } S(t)$$

$$H(t+1) = I(t) \text{ AND NOT } V(t)$$

$$I(t+1) = H(t)$$

$$S(t+1) = I(t)$$

It shows an oscillatory behaviour followed by a fixed point where the immune system is totally destroyed, similar to the real onset of AIDS.

A peculiar class of models in immunology is the so-called *bit-string* models. Farmer, Packard and Perelson [71] were the first to use bit-strings in a model of the idiotype network theory. All the molecules and cell binding sites (e.g., cell receptors) are modelled as binary strings of length l. Antibody molecules are assumed to recognise the antigen whenever their bit-strings can be matched complementarily. The specific rule that was used was to align the bit-string and require a complementary match over a stretch of at least r adjacent positions. For string matches over exactly r adjacent positions, a low affinity was assigned, say 10^{-1}. If the match was bigger than r adjacent positions, the affinity was increased.

The IMMSIM model has picked up the idea of bit-strings to simulate the match between an antigen's epitope and a lymphocyte's receptor. In the binary representation, the affinity between binding regions is a function of the number of complementary bits in the comparison (i.e., bit-strings have the same length). Take, for example, the cell receptor of a B lymphocyte that is represented by the binary string 00010101 ($l = 8$) and an antigen with an epitope represented by the string 11101010; the probability of triggering a response is a function of the match $m = 8$, hence it has maximum value. In order to be able to simulate affinity competition and maturation, some cases of mismatch are allowed to trigger the response, albeit with a lower probability. In other words, in this model, the bit-string match is not required to be perfect; some *mismatches* are allowed, for example, one bit out of eight. Each mismatch leads to a decrease in the strength of the affinity. Clearly, how to quantify this decrement is a modelling choice that can (and really does) influence the overall outcome in, for example, an immunisation experiment aiming to illustrate the affinity maturation phenomenon [82] discussed in section 1.19.

2.5 The Others

Besides C-ImmSim, there are a few other agent-based models of the immune system that should be mentioned. They have all been developed successively and perhaps the Celada-Seiden model inspired them. Here, they are briefly sketched.

One of the most interesting proposals is Simmune [83]; it aims at being a flexible platform to allow, in principle, the simulation of any process in immunology. Actually, it looks more a modelling technique and a language for the description of models than a specific model. Simmune is based on a particular representation of particle interactions that can be used to create detailed descriptions of the immune system. The particles live on a mesh and their states are (asynchronously) updated at discrete time-steps so that both time and space are discrete. Particles in Simmune have a continuous spectrum of states. Transitions between the states are probabilistic events triggered by the exchange of messenger particles with limited range. The messenger field intensities are calculated by integration of reaction-diffusion equations and typically include an activation threshold. In this way, Simmune models both direct inter-cellular interactions (such as those between an antigen and a B-cell) and interactions mediated by molecular messengers (such as lymphokines). It also supports spatial compartmentalisation and communication conduits. Overall, it provides a framework and a simulated environment for the immune system. As a matter of fact, the CS-model could be implemented by using the Simmune framework but this has never been attempted.

Also interesting is the Basic Immune Simulator (BIS) [84]. This model attempts to provide a complete simulation of the spatially extended dynamics in the thymus and to study the thymus selection by means of the agent-based paradigm. It details the interactions among the cells of the innate and adaptive immune system. The innate immunity, which is the initial response of the host to a pathogen, generally precedes the adaptive immunity, which generates immune memory for a specific antigen through a cascade of signalling and activations. The BIS simulates basic cell types, mediators and antibodies, and consists of three virtual spaces representing parenchymal tissue, secondary lymphoid tissue and the lymphatic/humoral circulation. The BIS effectively translates mechanistic cellular and molecular knowledge regarding the innate and adaptive immune responses, and reproduces the immune system's complex behavioural patterns. It can be used both as an educational tool to demonstrate the emergence of these patterns, and as a research tool to systematically identify potential targets for more effective treatment strategies for disease processes including hypersensitivity reactions (allergies, asthma), autoimmunity and cancer.

Simisys [85] is another simulator of the human immune system based on the agent-based paradigm. It allows the simulation of tens of thousands of cells as exemplars of the significant players in the functioning of the immune system, and reproduces normal and simple disease situations by interpreting interactions among the cells. Both innate and adaptive components of the immune system are represented. Specifically, macrophages, dendritic cells, neutrophils, natural killer cells, B-cells, T helper cells, complement proteins and pathogenic bacteria are present in the model.

Pathsim [86] is a spatial stochastic agent-based model constructed to study the immune response to Epstein-Barr virus

(EBV) infection. The model is based on an anatomically faithful representation of the Waldeyer's Ring, since the tonsillar ring is the primary site of infection and immune response. The blood and lymph compartments are represented as homogeneous reservoirs. The model consists of approximately 160,000 separate spatial locations made up of several different tissue types, and can presently be run with approximately 10^6 agents. Germinal centres are represented in detail, through 71 spatial locations. Agent types include virus, naïve B-cells, latently infected B-cells, as well as naïve and two types of activated T-cells. Agent motion is governed by diffusion, including diffusion that is sensitive to several different gradients. In addition to the simulator, the authors of Pathsim have developed a virtual environment that is navigable and employs a variety of information-rich virtual environment features to represent model output and navigate the simulation space [87].

There is another model or general methodology worth mentioning. It models biological systems as *reactive systems*. The authors, in [88], have constructed a model of T-cell reactivity in the thymus using "reactive animations" with the aim of solving the problem of heterogeneity, multi-scale modeling and the link between mathematical and computer models [89]. This methodology, extensively used in theoretical computer science and software engineering, describes the behaviour of heterogeneous entities by means of (deterministic or probabilistic) finite state automata, thus not altogether different from the solution adopted by C-ImmSim many years before (it will be described in the following sections).

By the time this book is printed there will be many more attempts to model the immune system by agents. The reason for this prediction is that, in the field of theoretical immunology, the ABM methodology, although not free from limitations, has become an essential technology for constructing *in silico* modeling architectures [90]. Nowadays, the interdisciplinary approach is perceived as a good choice in biology and medicine: quite a change since the time when IMMSIM was conceived.

3

C-ImmSim Unveiled

Immunology is a branch of biology that deals with the tasks performed by cells, molecules and organs, to protect the body of higher organisms such as fishes, reptiles, birds and mammals, against external assault of agents with potential ability to create a functional damage to the host itself resulting in a life threat. Almost any living organism evolved on the earth has a more or less sophisticated immune system. The most convenient view of the immune system is the one that identifies cells as the main actors of the immune response and the agent-based paradigm as the most straightforward way to implement it.

The C-ImmSim simulator, the direct successor of IMMSIM, will be described by first defining the actors and then their roles and the way they cooperate to build the organism's defence. These roles are disparate and intertwined. If the reader is an *alien*, and has gone through the first part of this book, by now, she/he should have at least a rough idea of the whole picture. However, those who are interested in getting deeper into the immune system could go for a textbook [91-93] or look at other bibliography resources [1, 44, 46, 94]. On the other hand, if the reader is a biologist and the words automaton, algorithm or data structure sound obscure to her/him, then it could be profitable to do some Internet search before reading further. Additional details about IMMSIM, C-ImmSim or the CS-model can be found in previous publications [1, 7, 82, 94-97].

Because C-ImmSim has evolved from the first version to account for different phenomena and to model various immune responses to pathogens, a separate section will describe some of its main modifications and upgrades.

As already mentioned earlier, the model includes both the *innate* and the *adaptive* immune response (see 1.43). Simply speaking, the innate immune response is the first defence and promptly eliminates 'minor' threats, whereas the adaptive immune response enters into action when the innate fails to recognise or to completely eliminate the invader. While the innate immune response in C-ImmSim is yet only marginally implemented by means of the function of the main antigen-presenting cells (macrophages and dendritic cells) and natural killers, the core of the adaptive immune response is fully implemented with lymphocytes B and T. The adaptive branch of the immune system is further specified in both *cellular* and *humoral immunity* (recall what was discussed in 1.43 and see Figure 23).

The whole adaptive immunity model hinges on the clonal selection theory of the Nobel laureate F. M. Burnet (1959) [9] developed on the pathways first traced by P. Ehrlich at the end of the nineteenth century. The clonal selection theory states that the *adaptive immune response* is the result of a selection of the 'right' antibody by the antigen itself (a fact that induces proliferation of the cells bearing the selected antibody), much like the fittest individual is selected by the environment in the Darwinian theory of natural selection. The requirements of clonal selection are that: (i) each immunocyte synthesises its individual, genetically coded antibody and exposes it on the outer membrane before and independently of the arrival of the antigen; and (ii) the selected cell grows into a clone allowing the production of the selected antibody to increase exponentially.

The CS-automaton is a *bit-string polyclonal lattice* model. Each of these terms has a specific meaning. In short, bit-string refers to the way the molecules and the specificity among molecules are represented, polyclonal indicates that more lymphocyte

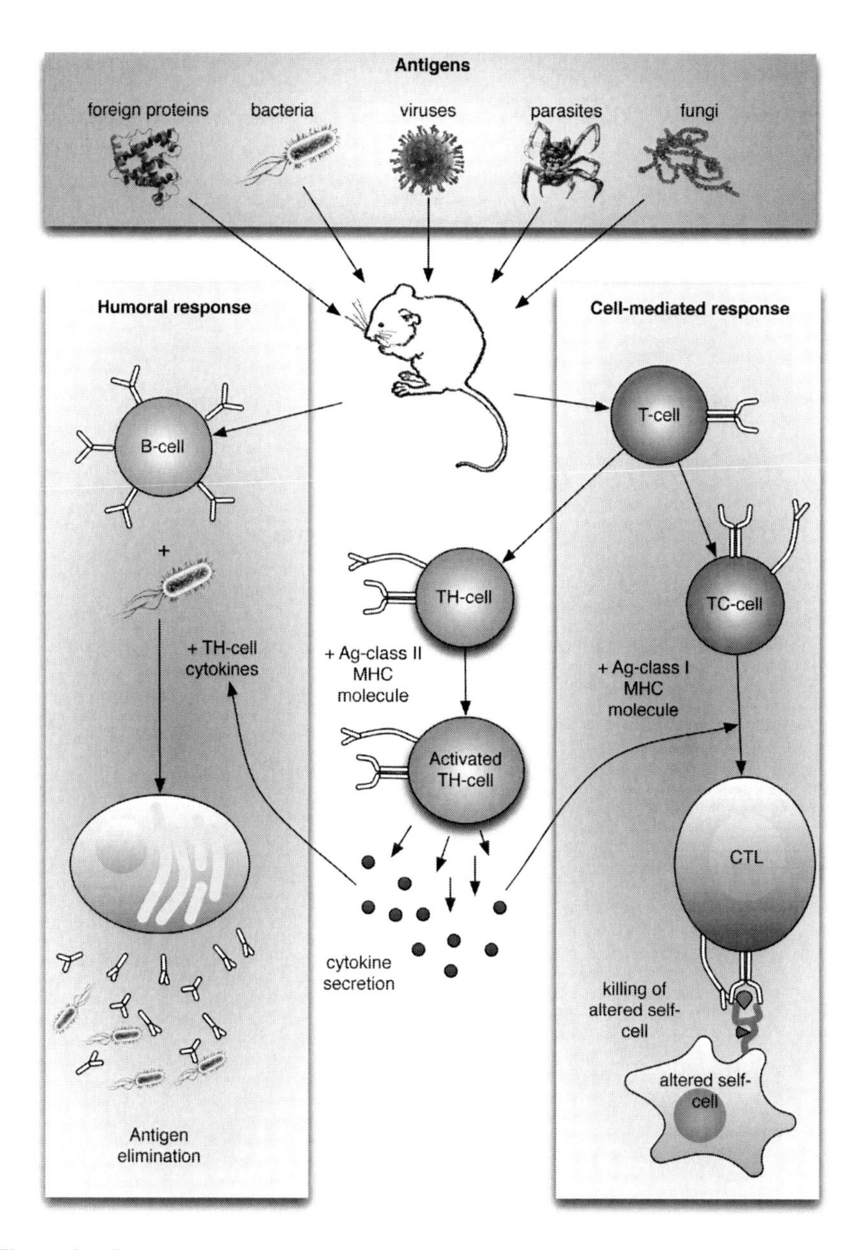

Figure 23. The two branches of the immune response to an offending antigen: humoral response, mediated by the production of antibodies, and the cellular response, mediated by the action of activated cytotoxic T lymphocytes. C-ImmSim implements both and enables the representation of various pathogens as virus and bacteria.

Colour image of this figure appears in the colour plate section at the end of the book.

clones of different specificity are represented (as opposed to the monoclonal models where only a single population of genetically identical lymphocytes is represented) and lattice means that it uses a discrete lattice to represent the space, that is, the space is discrete.

Major classes of cells of the lymphoid lineage (helper T lymphocytes and cytotoxic T lymphocytes; B lymphocytes and antibody-producing plasma B cells; natural killer cells) and some of the myeloid lineage (macrophages and dendritic cells) are represented in the model (listed in Table 3).

Being a model of a complex system, C-ImmSim includes different working assumptions, theories, hypotheses and known facts. Key points will therefore be described in greater detail below:

- Anatomical compartments
- Cells and molecules
- Repertoire
- Molecular affinity
- Haematopoiesis, generation of cells
- Cell maturation and thymus selection
- Hayflick limit in the number of duplications
- Aging and memory of past infections
- Hyper-mutation of antibodies
- Bystander effect
- Cell activation and anergy
- Cell interaction and cooperation
- Antigen digestion and presentation

Each of these subjects identifies a component or a process of the immune system. For the implementation of the specific functions, the CS-model refers to either a demonstrated or a hypothetical phenomenon. Each single component of the list above can, therefore, be implemented in a different way to test a new hypothesis or just to weigh the effects of a deviation on the system as a whole.

Table 3. Cellular and molecular entities and the symbol used in this book. Lymphocytes B, dendritic cells and macrophages are also called *Antigen-Presenting Cells* (APC) since they capture the antigen and, after a 'digestion' process, present it in small pieces to helper T lymphocytes.

CELLS	SYMBOL(S)
Lymphocyte B	**B**
Lymphocyte Plasma B-cell	**PLB**
Lymphocyte T helper, further subdivided as: Th1 Th2	**Th, CD4+ T** **Th1** **Th2**
Lymphocyte T regulatory	**Treg**
Lymphocyte T killer or cytotoxic cell	**Tc, CTL, CD8+ T**
Macrophage	**MA**
Dendritic cell	**Dc**
Natural killer	**Nk**
Epithelial or generic virus-target cell	**Ep**
Generic cancer cell	**Cc**

MOLECULES	SYMBOL
Immune complex or Ab-Ag binding	**IC**
Antigen (generic bacteria or Virus)	**Ag**
Generic antibody, further subdivided as: IgM IgG1 IgG2 IgE	**Ab** **IgM** **IgG1** **IgG2** **IgE**

CYTOKINES	SYMBOL(S)
Interleukin-2 Interleukin-12 Interferon gamma Interleukin-4 Transforming growth factor beta Tumour necrosis factor alpha Interleukin-10 Interleukin-6 Interferon beta Interleukin-18 Interleukin-23	**IL-2** **IL-12** **IFN-γ, IFN-g** **IL-4** **TGF-β, TGF-b** **TNF-α, TNF-a** **IL-10** **IL-6** **IFN-β, IFN-b** **IL-18** **IL-23**
Danger signal (generic)	**D**

3.1 Model Compartments

C-ImmSim is composed of three modules or compartments that correspond to three different and separate anatomical regions of mammals (see Figure 24): (i) the bone marrow where the haematopoietic stem cell is simulated by producing new lymphoid and myeloid cells; (ii) the thymus gland where T lymphocytes are selected to avoid auto-reactivity; and (iii) a tertiary organ like a lymph node.

In C-ImmSim, the space is discretised and a single lymph node of a vertebrate animal is mapped onto a two-dimensional hexagonal (or triangular) lattice, with periodic boundary conditions (Figure 25). A three-dimensional Cartesian mesh (six neighbours) can replace the two-dimensional lattice. Physical proximity is modelled through the concept of lattice-site. All interactions among cells and molecules take place within a lattice-site in a single time step, so that there is no correlation

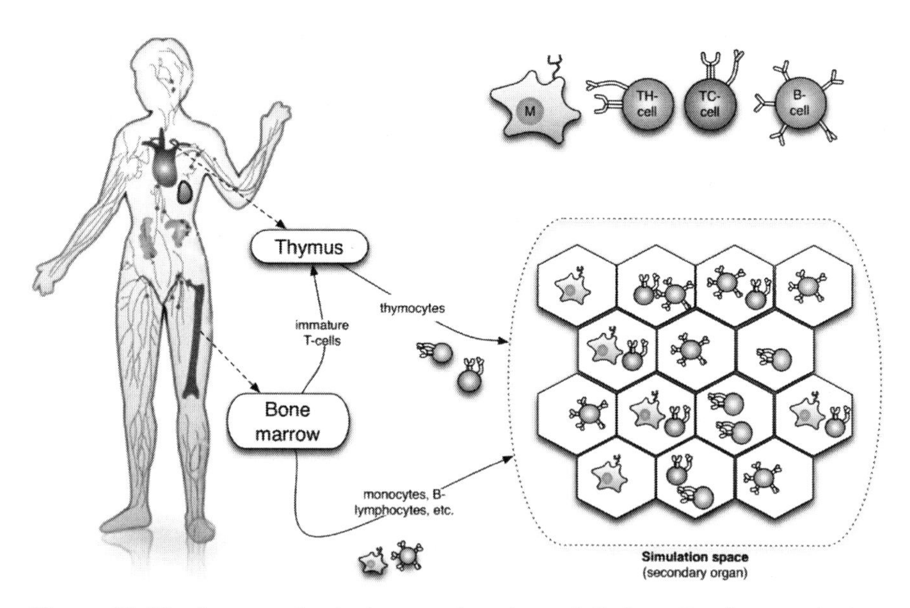

Figure 24. The three anatomical compartments modelled are the thymus, the bone marrow and a portion of a generic secondary organ (the picture of the human body is adapted from [98]).

Colour image of this figure appears in the colour plate section at the end of the book.

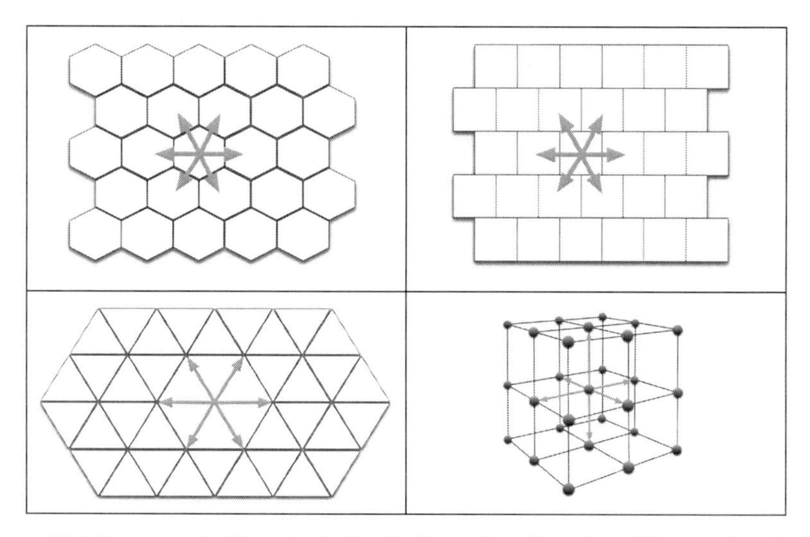

Figure 25. The space is discrete. In the bi-dimensional version of C-ImmSim the grid is a hexagonal lattice (top, left) or square-shifted (top, right). This is equivalent to the triangular lattice if you look at the edges instead of the nodes (bottom-left). In the three-dimensional version, the space is a Cartesian lattice (bottom-right).

Colour image of this figure appears in the colour plate section at the end of the book.

between entities residing in different sites at a fixed time. The diffusion of entities at the end of each time step is meant to model physical spreading of molecules. It introduces spatial relations, determines a time delay among events and influences the probability of favouring encounters.

Separate compartments simulate the bone marrow and the thymus. The thymus is implicitly modelled through positive and negative selection of immature thymocytes before they get into the lymphatic system (discussion on modules follows in the next section).

A set of self-peptides is needed to define the non-self (this is obtained, specifying the self peptides' binary strings in the problem's initial condition). The outside world of potential pathogens as well as the set of cell receptors and MHC Major Histocompatibility Complex molecules (the HLA or Human Leucocyte Antigen) is represented as a set of binary strings.

3.1.1 The Bone Marrow

The bone marrow is the place where the *multipotent* stem cells reside. Such cells differentiate into basically all kinds of haematopoietic cells according to a *plethora of stimuli*.

In the model, the differentiation is not taken into account since cells that are not ready to contribute to the immune response are, modelling-wise, irrelevant. Therefore, all cells are created and initialised in a state that is either 'active' or 'resting'. In both cases, they are mature and ready to do something. Once created, all cells reach the tertiary organ (i.e., the simulation space) and therefore they are mature and ready to be stimulated. T lymphocytes are exceptions, since before entering the circulation, they undergo selection in the thymus (section 1.17).

All lymphocytes generated in the bone marrow bear a receptor. The arrangement of the gene coding for the variable regions of the receptors is a quite complex issue that is matter of research. In the simple space of binary strings, the only reasonable choice, for an alien, is to generate the cell receptors for B and T-cells at random. Namely, given the bit-string length NBIT, and the function *rand*() that gives back a random integer between 0 and $2^{NBIT} - 1$, the receptor $rec = rand(0, 2^{\mathrm{NBIT}} - 1)$ is a binary string in its equivalent decimal form. Whereas the *potential repertoire* (that is the set of all possible receptors) covers the entire space $\{0, 1\}^{NBIT}$, the *expressed repertoire* depends on the number of cells, each carrying a single specificity, which are initialised in the simulation and will therefore be a subset of the potential repertoire. Clearly, since in reality the potential repertoire is much larger then the expressed one, the parameter NBIT, and the size of the simulated space L, which sets the number of cells to populate the simulation (see section 3.4), must be chosen with care.

3.1.2 The Thymus Gland

The *thymus* is the organ from which the T lymphocytes take their name. It is a major component of the lymphatic system. In the thymus, lymphoid cells undergo a process of maturation and education prior to release into circulation. This selective process allows the population of T-cells to develop self-tolerance (i.e., distinguishing self from non-self). While residing in the thymus gland, any T-cell that reacts to the self major histocompatibility complex (MHC) is eliminated. T-cells that tolerate the MHC learn to cooperate with cells expressing MHC molecules and are allowed to mature and leave the thymus. The result is that T lymphocytes surviving selection in the thymus tolerate the body's cells and cooperate with them when needed. However, in cases of imperfect selection, some T lymphocytes retaining anti-self aggressiveness may be released, which results in autoimmune diseases such as systemic lupus erythematosus or multiple sclerosis [91]. The thymus is implemented in C-ImmSim as a *filter* that, of the randomly generated receptors (i.e., cells bearing it), selects only those functionally useful (see 3.8). A modified version of IMMSIM [96] has been constructed to simulate the thymus in a more detailed way also taking into account the volume. A similar study using a modified version of C-ImmSim accounting for the molecular primary structures has produced similar outcomes [99].

3.1.3 The Lymph Node

The model represents a *portion* of a secondary organ like a lymph node, the tonsil, the spleen, Peyer's patches (in the small intestines), and the appendix. Secondary organs are the sites where lymphocytes localise and identify unfamiliar antigens and trigger a reaction in opposition to them. This compartment is also described geometrically, in contrast to the bone marrow and the thymus. However the geometry is quite simplistic and just the notion of physical proximity is modelled at the current stage. Although it is possible in theory to represent the lymphatic system in its entirety (i.e., lymph nodes and channels), in practice there is a limit imposed by the computational requirements. In fact, although it would not be pointless to simulate realistic immune cell trafficking in anatomical compartments and through lymph nodes and channels, to date there has not been a real attempt to challenge this topic by the modelling community.

3.2 The Cells

Since C-ImmSim describes the world at the cellular level, the cells are the key actors and need to be well defined and represented. From the data structure point of view, the cells are just records or objects (in the Object Oriented Paradigm). In abstract, the cells can be described as *finite state machines*. In particular, they can take up a state from a certain set of suitable states and their dynamics is realised by means of state changes. A state change takes place when a cell interacts with another cell or with a molecule. In this way, one focuses on the logic of the network of available interactions, which resemble, although in a much more complicated way, the model of Kaufman, Urbain and Thomas [65]. Therefore, in C-ImmSim, the theoretical model of the network of interactions is realised by determining all the possible states for each biological entity represented, and then writing down the interactions in terms of (i) the state which characterises the cells that can interact (conditions) and (ii) the new state of the cell after the interaction (actions or post-state).

By using a mathematical formalism, one can write each interaction as a transformation of this kind:

$$(S(e_1), S(e_2), t + 1) = I(S(e_1), S(e_2), t)$$

$\forall (e_1, e_2) \in E \times E$, where $S(e)$ identifies the state of the entity $e \in E$, E is the set of entities, and $I : S \times S \times T \mapsto S \times S$ is the function determining the state at the following time step t given the internal state of the entities. To be more precise, since the definition of a state transition procedure does not change with time, one should write $I : S \times S \mapsto S \times S$ and

$$(S(e_1), S(e_2)) = I(S(e_1), S(e_2))$$

to indicate the fact that the transition is determined solely by the state of the entities, or in other words, that the fate of a cell depends upon what it actually is (i.e., the activation state or maturation stage). Mathematicians would call this cellular dynamics a Markov process.

3.2.1 Entity-state Description

The model entities are either cells or molecules. From the point of view of a computer programmer, the former need to be represented by a much more sophisticated data structure because of their inherent higher complexity. In particular, cells have an internal dynamics, which translates as a variable to store the internal state, one for each instance. Moreover, each lymphocyte is equipped with a receptor described as a binary string of a certain length. In contrast, antigen and antibody molecules can be distinguished just by the binary representation of the *epitope/peptide* (for the antigen) and the *paratope* (Fab, for the antibody, see Figure 33). Since for a given binary configuration, the molecules are exactly the same, there is no need to describe them with an internal state. In other words, molecules are not represented as agents. Lastly, small molecules such as interleukins and the like are practically indistinguishable, and therefore can be all taken together in a single variable which stands for the total number of molecules (or the molecular concentration) in each lattice site (position).

Since each cellular entity has a state and the transitions among states determine its behaviour, the single cell can be considered as a *Stochastic Finite State Machine* or Automaton (SFSM [100]), which processes information and changes its state according to the results of the interaction with other cells. The transitions among the various states are determined by stochastic events. The set of possible transitions of the SFSM's is defined by the entity-state description reported in Table 4.

Table 4. At each time point, a cell can be found in one of the states reported in the first column if an X marks the corresponding entry in the table. Note that PLB is not shown in this table; it does not really need an internal state since it is committed to produce antibodies during its entire lifetime. Th1, Th2 and Treg all have the same state definitions as the Th.

State	Description	B	Th	Tc	MA	Dc	Nk
ACTIVE	In this state, the cell is mature and ready to interact. This is the initial state for most cells.	X	X	X	X	X	X
ANERGIC	This is the anergic state. In this state, the cell does not interact.	X	X	X			
INTERNALISED	In this state, the cell has engulfed one antigen.	X			X	X	
PRESENTING-1	MHC-1 molecule is loaded with one antigen peptide.					X	
PRESENTING-2	The cell has processed the antigen it previously engulfed. The cell is now exposing the MHC-2 molecule bond with an antigen peptide.	X			X	X	
DUPLICATING	The cell is duplicating. It will remain in this state for a number of duplication steps.	X	X	X			
RESTING	The cell is in the resting state, waiting for a signal to be activated.		X	X	X		X
SILENTLY INFECTED	A virus has infected the cell and its DNA is already part of the cellular genome. The virus is not replicating.					X	
ACTIVELY INFECTED	The virus that has infected the cell is actively replicating.					X	
DEAD	The cell has been marked to die by lysis by a cytotoxic cell or is necrotic for other reasons.	X	X		X	X	

All other entities (i.e., the molecules) may be considered always ACTIVE, in other words, ready to interact. Note that specifying the behaviour of entities in terms of discrete states is an advantage in terms of memory parsimony since the cell state is stored in a *flag* byte for each cell. Specific *macro-instructions* have been defined to access, both in read and write mode, the flag byte. This access is also computationally very cheap.

Conceptualising the dynamic evolution of biological systems in terms of state-transitions of biological objects is quite conventional in immunology and among biologists at large. Therefore, adopting such a description to describe the behaviour of cells is definitely an advantage. That is one reason why this idea has been recently expanded by adopting the Unified Modelling Language state-transition diagram description. Another not insignificant reason to use UML is to ease software coding, manipulating and documenting [89].

3.2.2 The B Lymphocyte

The first description in terms of stochastic finite state machines is that of B-cells. The way of interpreting it is minimalistic. B-cells start in the ACTIVE state. This state's shape is colour coded in grey in Figure 26. An arrow labelled with a few explanatory

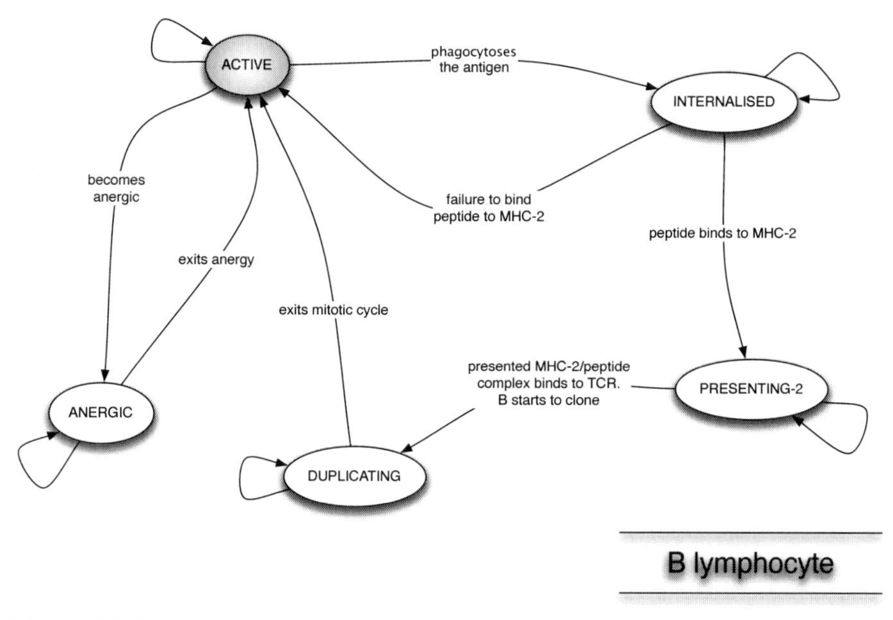

Figure 26. B-cells start in the ACTIVE state. They can go back and forth between ACTIVE and ANERGIC according to the stimulation received. While ACTIVE, if they engulf an antigen, they go to the INTERNALISED state. If the MHC-2 succeeds in binding one antigen peptide, the B-cell goes into the PRESENTING-2 state. A cell in this state is able to stimulate a matching T helper lymphocyte and begin a clone expansion.

words indicates a transition to another state. Each transition corresponds to a logical rule of the algorithm. If the B-cell receives a massive antigenic stimulation, it goes to ANERGIC (see section 3.14). Since anergy is a kind of temporary refractory state, the cell can jump back to the ACTIVE state with a certain fixed probability. In the ACTIVE state, the cell is able to engulf an antigen. Upon successful phagocytosis of the antigen, it goes to the INTERNALISED state. From there, it can go back to ACTIVE after an unsuccessful digestion of the antigen or it can go to PRESENTING-2 state if one 'piece' of the antigen (the *peptide*) binds to the MHC-2 molecule. In this state, the B-cell exposes the molecular complex formed by the MHC-2 molecule and the free end of the peptide (in C-ImmSim a peptide is half bit-string, so when one half binds to the MHC-2, the other half is the piece that is exposed together with the MHC-2 on the cell surface; see Figure 42). When in PRESENTING-2, the cell may interact with a T helper lymphocyte through the binding of the MHC-2/peptide complex with the T-cell receptor. If this happens, the state changes into DUPLICATING, and the cell starts to clone. It will remain in this state for a certain number of cycles. After that, the B-cell goes back into the ACTIVE state, ready to engulf another antigen and start the whole process again.

3.2.3 The Macrophage

The macrophage (Figure 27) starts in the RESTING state but can go into the ACTIVE state according to the amount of danger signals (see section 3.13.1). The activation is actually indirect since natural killers will secrete IFN-γ in response to danger and IFN-γ will drive the state change from RESTING to ACTIVE. On the other hand, in the absence of stimulation, a macrophage tends to stay in the RESTING state. Thus, with some probability, it goes from ACTIVE back to the RESTING state. Active macrophages can get to the INTERNALISED state upon successful phagocytosis of an antigen molecule. After that, as in the case of the B-cell, the MA can go to the PRESENTING-2 state or back into the ACTIVE state depending

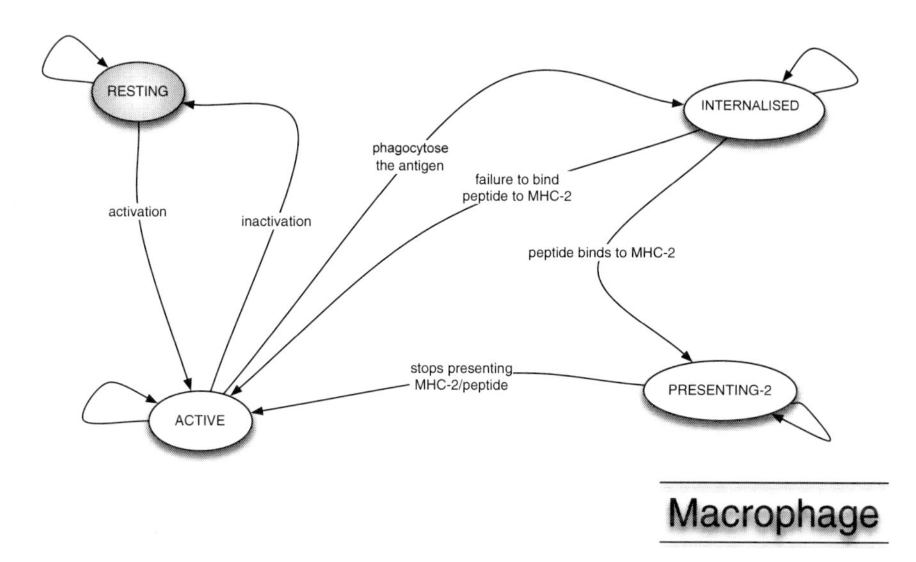

Figure 27. The macrophage starts in the RESTING state but goes into the ACTIVE state with a certain probability that depends on the level of the interferon-γ signal. Once active, a macrophage swallows, digests and presents the antigen's peptides to the helper cells in the same way that the B-cell does.

on the success of the matching between the peptides composing the phagocytosed antigen and the MHC-2 molecule. In contrast to the B-cell, the macrophage does not duplicate and remains in the PRESENTING-2 state until it interacts with a T helper lymphocyte or it stops presenting the MHC-2/peptide. Then it goes back into the ACTIVE state to start the cycle all over again.

3.2.4 The Helper T Lymphocyte

Helper T-cells (in Figure 28) start in the RESTING state and become ACTIVE when their membrane receptors come in contact with an MHC-2/peptide complex on the cell surface of macrophages or dendritic cells (i.e., those which are in PRESENTING-2 state). If a T-cell encounters a B-cell presenting the antigen peptide before it gets the priming signal from macrophages or dendritic cells, it becomes ANERGIC. Instead, once active, T-cells can be stimulated again by any antigen-presenting cell bearing the MHC-2/peptide complex. This stimulation starts the mitotic cycle and the B-cell and its progeny enter the state,

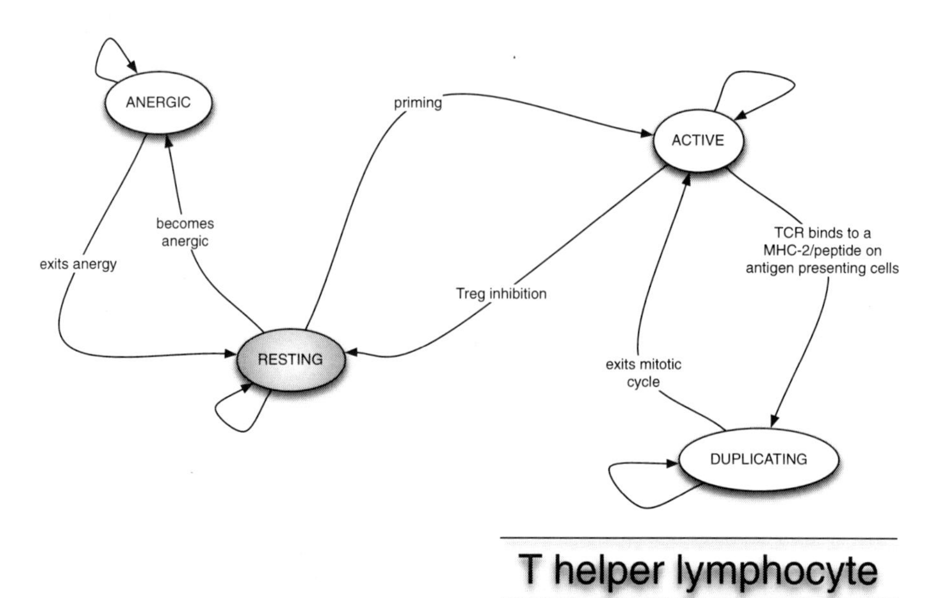

Figure 28. T helper cells start in the RESTING state and can become anergic if a B presenting cell stimulates them before they get the priming signal by macrophages or dendritic cells. They become ACTIVE if stimulated by macrophages or dendritic cells instead (these cells have to carry the MHC-2/peptide complex on their surface). Once ACTIVE, they can be stimulated again by any antigen-presenting cell bearing the MHC-2/peptide complex. Upon stimulation, they go to the DUPLICATING state (the mitotic cycle). After completion of the duplication phase, they go back to the ACTIVE state again. In the ACTIVE state, they can be inhibited by T regulatory cells and enter the RESTING state again.

DUPLICATING. After completion of the duplication phase, these cells return to the ACTIVE state. Note that, once activated, the helper T-cell does not go back to the RESTING or ANERGIC state but remains committed to be active or duplicating, unless it interacts with a T regulatory specific cell.

The model includes two subclasses of helpers: Th1 and Th2. The same stochastic finite state machine, as in Figure 28, can describe their behaviour. Regulatory T-cells can be as well illustrated by this SFSM with the only difference that they are inhibited by nothing.

3.2.5 The Cytotoxic T Lymphocyte

The CTL starts in the RESTING state (Figure 29). It must bind the MHC-1/peptide complex on target cells (for instance epithelial cells) and after getting IL-2 from helper T-cells (priming), it maturates into an effector cell (equivalent to the state ACTIVE). ACTIVE specific CTLs can kill cells and be stimulated to duplicate. After the duplication phase, the cytotoxic T-cell returns to the ACTIVE state. Once active, the cell is committed to stay in the ACTIVE state until it dies of aging or, upon interaction with a T regulatory specific cell, it enters the RESTING state again.

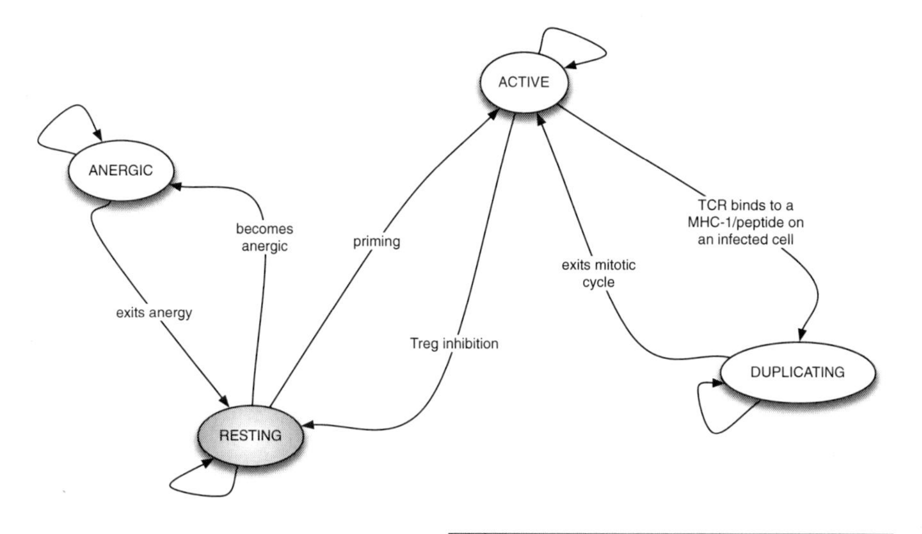

Figure 29. Cytotoxic T lymphocytes start in the RESTING state. They must bind MHC-1 on target cells (e.g., epithelial cells) and get IL-2 from T helper lymphocytes to maturate into effector active cells. ACTIVE CTLs can kill infected or malignant cells presenting a peptide on MHC-1 and be stimulated to duplicate. After the duplication phase, the cytotoxic T-cell goes back into the ACTIVE state. Active T cells can be inhibited by T regulatory cells and enter the RESTING state.

3.2.6 The Dendritic Cell

Dendritic cells are very efficient antigen-presenting cells. Their shape, with long arms, makes them very capable of capturing an antigen. Dendritic cells (look at Figure 30) start in the RESTING state and can be activated by an excess of the inflammatory cytokine TNF-α released by macrophages. Once activated, and upon successful internalisation of the antigen, they go to the INTERNALISED state. From there, the dendritic cell can digest the antigen peptides and present them with the MHC-2 molecule to the T helpers (state PRESENTING-2). Then, with a certain fixed probability, they can stop presenting the peptide and go back to the ACTIVE state to start the cycle again. An active dendritic cell can become inactive (go to the RESTING state again) if the level of 'inflammation' represented by the TNFα concentration falls below a certain threshold.

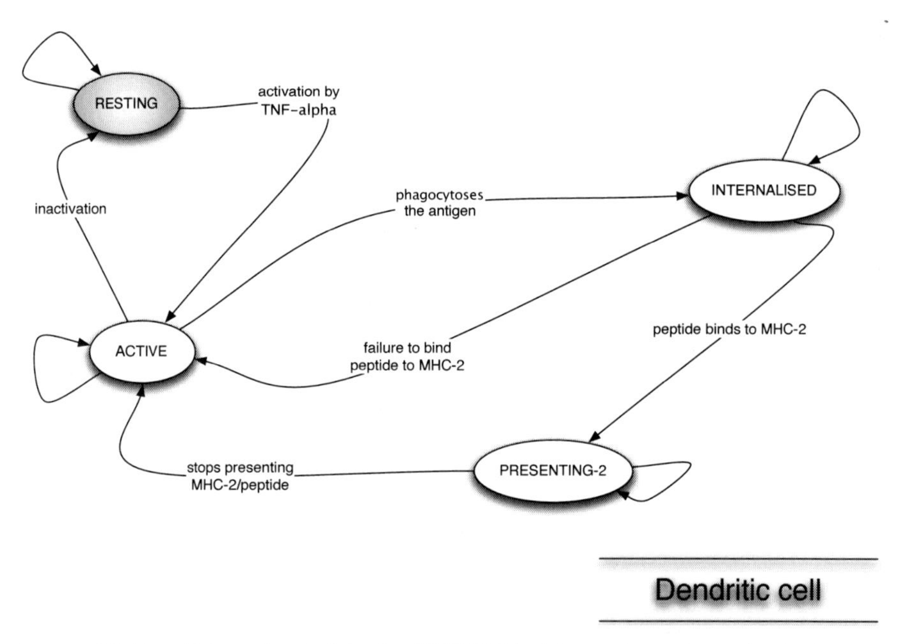

Figure 30. Dendritic cells start in the RESTING state and can be activated by an excess of the inflammatory cytokine TNFα, released by MA. From the ACTIVE state, upon successful internalisation of the antigen, they go to the INTERNALISED state. From there, the dendritic can digest the antigen peptides and present them with the MHC-2 molecule (state PRESENTING-2) to T helpers. Then, with a certain fixed probability they can go back to the ACTIVE state and, in absence of TNFα, back to the RESTING state ready to start the cycle again.

3.2.7 Natural Killers

Natural killers' behaviour is quite simply represented (Figure 31). They are initially RESTING and are activated by generic danger signals (e.g., uric acid) released, for example, by damaged cells or cells that are dying. Active natural killer cells are committed to patrol the tissues for their entire lives.

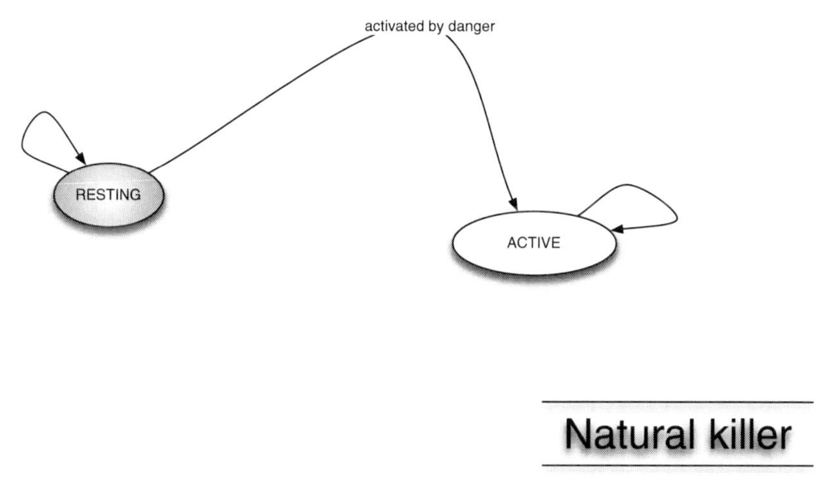

Figure 31. Very simple dynamics of natural killer cells. They are initially RESTING and are activated by generic danger signals. Upon activation, they release IFN-γ which, in turn, activates macrophages.

3.2.8 The Epithelial (or Generic Virus Target) Cell

Epithelial cells represent generic virus target cells. As such, they have no function other than staying in the ACTIVE state, ready to be infected by a virus (Figure 32). An epithelial cell whose genome embodies the viral DNA is said to be in the SILENTLY INFECTED state. From this state, the cell goes to the ACTIVELY INFECTED state if the virus is assembled, and hence replicates. Depending on the type of virus considered, either one of the following two mechanisms of viral replication operates: (i) the cell in the ACTIVELY INFECTED state releases copies of the virus budding from the membrane; (ii) when viral copies inside the cell reach a certain threshold, the cell explodes and

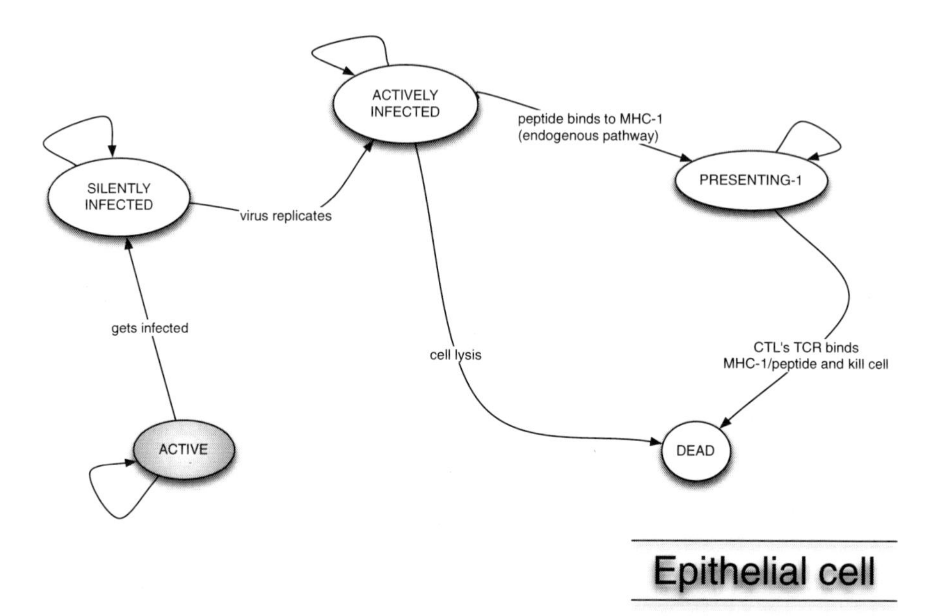

Figure 32. ACTIVE epithelial cells are targets of the virus. A virus first integrates its DNA into the cell's DNA and later, upon activation, it starts replicating. This is when the epithelial cell changes from the state of SILENTLY to ACTIVELY INFECTED. In the latter state, the cell may release virus particles budding from the cell membrane. The fate of this cell is signed and it will die either because its membrane will break when its viral content reaches a threshold or because a cytotoxic T lymphocyte kills it. This happens if the TCR of the cytotoxic T binds to the MHC-1/peptide complex exposed on the infected cell membrane upon digestion of the viral peptides.

releases its viral content in the surroundings. In the ACTIVELY INFECTED state, however, the epithelial cell is able to digest the virus and present the viral peptides with the MHC-1 molecule to T cytotoxic cells (PRESENTING-1). When a CTL recognises the MHC-1/peptide on the membrane of an infected epithelial cell, it kills the epithelial cell (DEAD). A dead cell is in the necrotic state, about to be eliminated through cytolysis.

3.3 The Molecules

The model distinguishes between signalling molecules like cytokines or interleukins (for them, it is sufficient to have a measure of the concentration on each lattice point) and more 'complex' molecules like immunoglobulins that need to be represented with a binary string.

In nature, immunoglobulins are divided into five different classes based on differences in the amino acid sequence in the constant region of the heavy chains [101]. All immunoglobulins within a given class will have very similar heavy chain constant regions: IgG-γ, IgM-μ, IgA-α, IgD-δ and IgE-ε heavy chains. In general, however, IgG is prominent and the most versatile immunoglobulin, capable of carrying out all the functions of immunoglobulins. For the purpose of simulating its effect in a virtual immune system, suffice it to say that IgG is the major immunoglobulin in serum (75% of serum antibodies) and the major immunoglobulin in extra vascular spaces. This is the reason why most mathematical or computational models do not make the distinction between the five classes of immunoglobulins. For example, in IMMSIM, the immunoglobulins represented are the IgG type.

In C-ImmSim, instead, immunoglobulins are differentiated in order to account for specific processes, e.g., hypersensitive type I reaction (allergies). In this case, in fact, it is crucial to distinguish between IgE and other immunoglobulins since IgE are responsible for the sensitisation of the granulocytes, which release the histamine and other cytokines [102]. Besides, the distinction between IgG1 and IgG2 subtypes is important when simulating the infection by the influenza virus. This differentiation phenomenon, called 'isotype switch', has been implemented on the basis of suitable rules and added to the model to simulate the infection by H1N1 [103].

As far as the antigen is concerned, it is a generic molecule with a certain degree of *immunogenicity* (i.e., the ability to induce an immune response) or a more complex structure, like virus and bacteria. Whatever it is, the antigen is represented by a collection of binary strings (the pieces), some representing what is exhibited on the *outside* and available for binding by receptors, and some which are not available, because they are

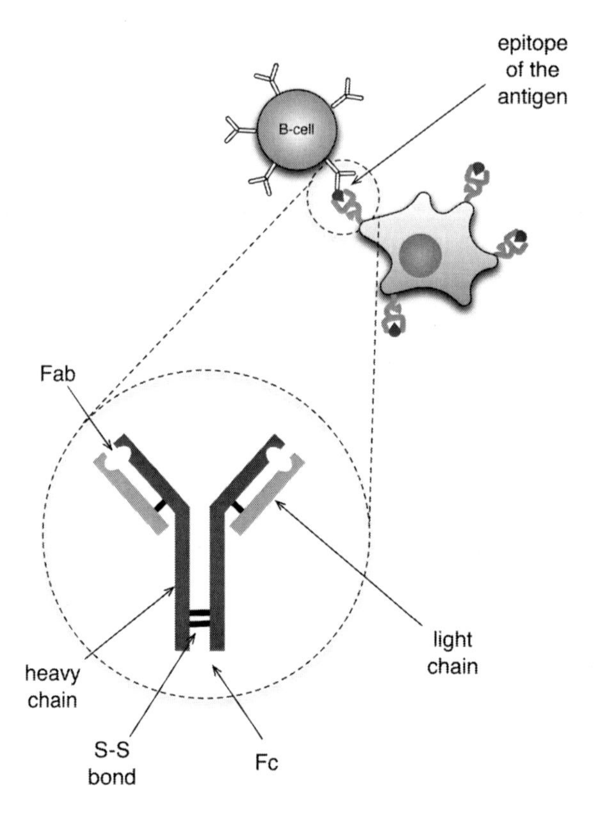

Figure 33. The major type of immunoglobulin, IgG, is the one that is often used to represent the antibodies in C-ImmSim and the only one used in IMMSIM while a subdivision in classes IgG1, IgG2 or IgE has been done to account for significant processes in specific diseases like hypersensitivity (section 5.1) or flu.

Colour image of this figure appears in the colour plate section at the end of the book.

hidden inside. Therefore the antigen is a collection of N_{inside} binary strings called the *peptides* and $N_{outside}$ binary strings called *epitopes*. While the epitopes account for the immune recognition by the B-cell receptor as discussed in section 3.5, the peptides are taken into account during antigen digestion and presentation on histocompatibility complex molecules, which will be explained in section 3.16.

3.4 The Repertoire

A clone of cells is a set of cells of the same type that show the same *receptor* on the cell surface. In the CS-model, receptors are represented by binary strings. The length of the binary strings NBIT is clearly one of the key parameters in determining the computational complexity (both time and space) of the simulator as is the number of potential repertoire of receptors scales, 2^{NBIT} (see [94]).

The repertoire of a biological entity (agent) is then defined as the cardinality of the set of possible instances of entities that differ in, at least, one bit of the whole set of binary strings used to represent all its molecules. It follows that the cells equipped with binding sites (i.e., the lymphocytes), as well as the antibodies, have a potential repertoire of $2^{N_e \cdot \text{NBIT}}$, where N_e indicates the number of binary strings used to represent receptors, MHC/ peptide complexes, epitopes and so on, of the entity *e*. Other entities do not need to be specified by binary strings so their repertoire is just one (i.e., $N_e = 0$). Examples are the interleukin molecules and the danger signal.

Table 5 summarises the number of strings used to represent each entity. Since the number of different MHC molecules (class-1 or 2) is limited to a few units, they do not contribute to the complexity of the cells (all cells equipped with MHC molecules carry the same number of the same MHC molecules). Actually, what one needs to represent specifically for each cellular entity is the MHC/peptide complex produced by the internal processing of antigens (*endocytosis*). This bit-string information is important because it is used for further recognition of the T (helper or killer) lymphocytes.

3.4.1 The Ratio Between Expressed and Potential Repertoire

It has been estimated [46] that the potential number of different lymphocyte receptors is $10^{11} \sim 2^{36}$ for B-cells and $10^{13} \sim 2^{43}$ for T-cells, which is far beyond reach for a binary representation

Table 5. Each cellular entity is characterised by its repertoire. For example, $N_e = 2$ for B-cells because each of them carries a receptor and an MHC-2/peptide molecule. For IC and Ag, N_e is the number of strings chosen to represent the epitopes ($N_{outside}$) plus those used for the peptides (N_{inside}). Cytokines like IL-2 and the danger signal do not carry any specificity, hence they have a repertoire of 1. Epithelial and cancer cells have a repertoire of 2^{NBIT} because they can express an endogenous peptide on class-1 MHC molecule.

Entity		**What**	**Repertoire**
B	2	Receptor and MHC-2	$2^{2\times NBIT}$
MA, Dc	1	MHC-2	$2^{NBIT}2^{2l}$
Th, Tc	1	Receptor	2^{NBIT}
PLB, Ab	1	Receptor	2^{NBIT}
Ep, Cc	1	MHC-1	2^{NBIT}
IC, Ag	$N_{inside} + N_{outside}$	Epitopes and peptides	$2^{(N_{inside} + N_{outside})\times NBIT}$
Nk	0	Unspecific receptor	2^0
Interleukins, cytokines, etc.	0	Signal	2^0

in current computers. This means that each *in silico* experiment should be performed keeping in mind that one is limited to a considerably small set of possibilities and this may (or may not, according to the kind of experiments done) influence the conclusions. To better understand this fact, it is worth defining the difference between the expressed and the potential repertoire.

The *expressed* repertoire of B-cells (i.e., the coverage of the potential set of receptors at any time), starting with a binary string of length NBIT and minimum match m_0, is

$$B^{expr}(0) = \frac{1}{2^{NBIT}} \sum_{m=m_0}^{NBIT} \binom{NBIT}{m}$$

For example, for NBIT = 12 and $m_0 = 9$, $B^{expr}(0) \cong 0.073$. This value, together with the cell and antigen density, determines the time it takes the immune system to recognise the antigen and to mount a humoral immune response. Note that it is not taken into account that Th recognition of MHC/peptide is necessary for the immune system, and that this is dependent on the definition of self-molecules and on the efficiency of the thymus selection.

The time taken to mount an immune response and the time it takes to get rid of the antigen is a way to determine the value of some parameters of the simulation. In particular, the probability of recognising the antigen depends on the expressed repertoire and also on the concentration of both cells and antigens in the simulated space. Therefore, it is possible to 'rescale the system' so as to obtain realistic delays in the advance of the immune response.

3.5 The Molecular Affinity

The bit-string model uses a string of bits (0s and 1s) to represent specific elements or binding properties in the same way that Farmer, Packard and Perelson did [71]. Each different bit-string defines an element of the *repertoire*. The greater the number of bits NBIT, the more diverse are the specific elements and thus the available specificities, the repertoire. Two bit-strings complement each other (or are a perfect match) if every 0 in one bit corresponds to a 1 in the other and conversely. More generally, an m-bit match is obtained when exactly m bits complement each other and the other NBIT – m are equal.

The function match(a, b) = hamming(a, b) is defined to give us the number of matching bits between two strings a and b and is computed as the Hamming distance in the space of the bit-strings (Figure 34). Similarly, mismatch(a, b) = NBIT – match(a, b) gives the number of non-matching bits. The function Affinity(m) is defined to give the affinity of an m-bit match. Typically, one wants to severely penalise mismatches to ensure that the perfect matches prevail over imperfect ones.

The function Affinity(m) is defined by a vector of length bits, called bit match vector, with each component of the vector giving the affinity of an m-bit match. To specify the vector Affinity, one method is to specify it directly by simply listing out the components of the vector. Another method uses the

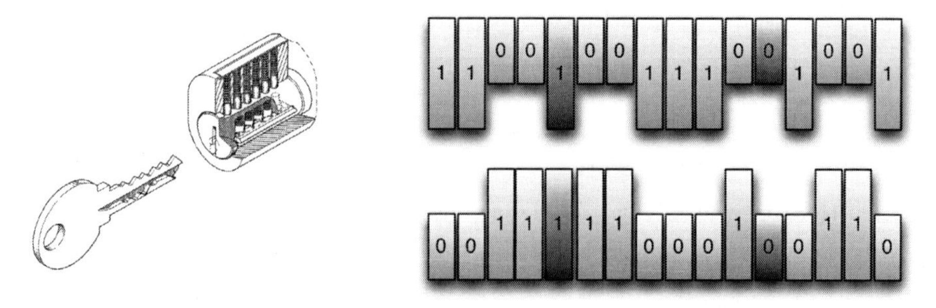

Figure 34. The match between two strings reminds one of that between a lock and a key. It is computed as the Hamming distance in the space of all possible binary strings of a certain length. In this example, 16 bits are used. The mismatch is two because two pins (those in red) do not pair.

Colour image of this figure appears in the colour plate section at the end of the book.

additional parameters, m_0, A_0, ∂A to calculate the vector in the following way: (i) first, set Affinity(m) = 0 for $m < m_0$; this provides a threshold level below which binding cannot occur; (ii) set Affinity(m_0) to the parameter A_0; (iii) set the increase of strength on increasing a match by one bit to be the inverse of the ratio of number of clones with match $m + 1$ and m multiplied by ∂A.

$$\frac{\text{Affinity }(m-1)}{\text{Affinity }(m)} = \partial A \cdot \frac{\binom{\text{NBIT}}{m}}{\binom{\text{NBIT}}{m-1}}.$$

This definition allows setting the lower end value of Affinity(m) and the steepness of its increase, as the number of matching bits is incremented (Figure 35). It is usually more convenient than supplying the Affinity vector directly. Generally, it is advisable to set a minimum match m_0 somewhat close to NBIT bits in order to restrict the range of allowed matches to few bits. In doing so, the rise in the number of antibodies in response to a given antigen remains manageable.

The affinity among MHC molecules (both class-1 and 2) and the peptides (endogenous or exogenous) is computed in a slightly different manner. Firstly, the match is a computed over

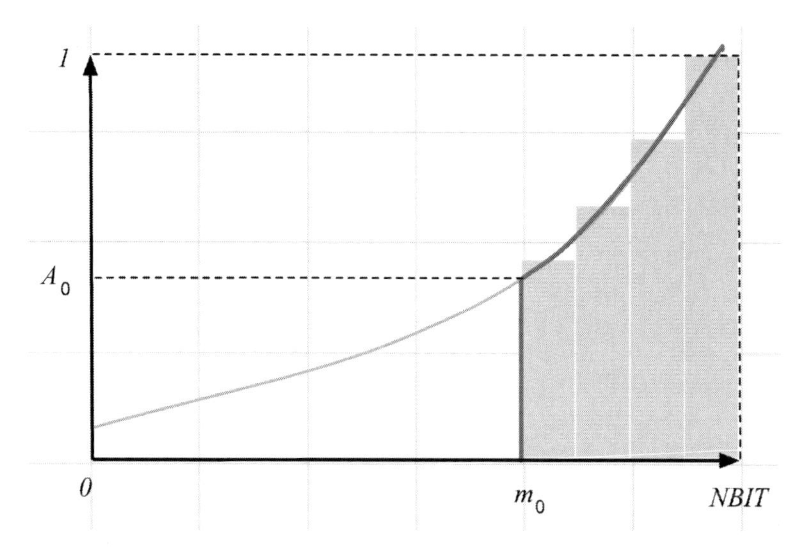

Figure 35. Affinity function. The probability to bind is a function of the Hamming distance between the two binding sites of the interacting entities. A match of m_0 or greater is required to have positive affinity.

Colour image of this figure appears in the colour plate section at the end of the book.

half bit-string.Secondly, there is no minimum matching. The affinity value between two half strings whose match is m, $\forall m = 0 \ldots \text{NBIT}/2$,

$$\text{MHCAffinity}(m) = 2^{(m - \text{NBIT}/2)}$$

This function is used in the antigen digestion by antigen-processing cells as explained in the forthcoming paragraph 3.16.

3.6 Reshaping the Affinity Landscape

The recognition of antigenic "signatures" (i.e., epitopes and peptides) is a *pattern recognition* ability that is mediated by mechanism, mostly unknown in details, based on different physical effects (short range noncovalent interactions, hydrogen binding, van der Waals interactions, etc. [46]). The concept of *generalised shape* that has already been introduced at the beginning of section 2.4 allows the identification of the array of features important in determining the binding among molecules. It follows that the repertoire of an individual consists of a set of generalised shapes. In other words, at any one time in the life of the individual, the 'statistical' distribution of the generalised shapes can be extended to all kind of receptors (e.g., PAMPS, etc.), in order to account also for the unvaried innate immune receptor repertoire. What results is a *multi-dimensional surface*, generically called immunological landscape. The affinity potential can thus be defined as the degree of affinity of an antigen to each point in the immunological landscape.

A naïve immune system has a 'flat' immune landscape because the generation of receptors is random in nature. Subsequent encounters with antigens elicit immune responses and reshape the landscape because the population of recognising clones expands at the expense of other, less useful clones (the next step is the MaN effect described in section 1.50). Because of this hustle the landscape starts to show hills and valleys and the advent of memory responses enhances all differences. The memory itself *is in* the hills and valleys (see Figure 36).

Figure 36. The immunological landscape is reshaped by endogenous (tolerance mechanisms) and by exogenous (antigenic challenges) stimuli.

The concept of shape space has been exploited for a long time in many mathematical models of the immune response. C-ImmSim is one of such models, in which the idea of immunological landscape as a 'learning' process of the immune system has been supported. By recognising the antigen epitopes and building the immunological memory, the immune system actually 'learns' the shape of the antigen [82]. The landscape curvature is how the system represents the stored *information* that, in other words, stands for its vision of the external world.

The multi-dimensional surface that identifies the immunological landscape at time t, can be described as $M(t; H; E)$, that is, not just a function of time t, but also of the history H, and the environment, E. This notion represents the immune system's vision of the world as an ever-changing reality, and its necessary response, a continuous adaptation and coevolution with the environmental stimuli. This condition, in general terms, is the Darwinian preferred choice in a mutable habitat.

3.7 Haematopoiesis and Cell Homeostasis

Haematopoiesis is the formation of blood cells derived from haematopoietic stem cells. In the model this process includes production of B and T lymphocytes in the "bone marrow compartment" and further selection of Th and Tc cells in the "thymus compartment". In order to keep the system in a metastable state in absence of perturbations (cell *homeostasis*), haematopoiesis is modelled as an Ornstein-Uhlenbeck mean reverting process [104]. The mean reverting process is a stochastic process that describes the velocity of a Brownian particle under the influence of friction. Over time, the process tends to drift towards its *long-term* mean. That is why it is called mean reverting.

More in details, given by the stochastic differential equation for the number of cells:

$$dx(t) = \eta(\bar{x} - x(t)) + \sigma d\varepsilon(t),\ x(0) = x_0$$

which means that in this case,

$$\frac{dx_t}{dt}(t) = \frac{ln2}{\tau_i}(x_i(0) - x_i(t)) + \frac{d\varepsilon}{dt}(t),$$

for constants $\bar{x}$, and x_0 and where $\varepsilon(t)$ is standard Brownian motion. In this model the process $x(t)$ fluctuates randomly, but tends to revert to some constitutional level $\bar{x}$ which here corresponds to the initial value x_0. The behaviour of this 'reversion' depends on the short-term standard deviation σ and on the speed of reversion parameter $ln2/\tau_i$. The Ornstein-Uhlenbeck process is a Gaussian model in the sense that, given x_0, the value of the process $x(t)$ is normally distributed with mean $E[x(t) \mid x_0] = \bar{x} + (x_0 - \bar{x})e^{-nt}$ and variance $Var[x(t) \mid x_0] = (\sigma^2/2n)(1 - e^{-2\eta t})$, where η characterises the propensity to return to the mean value. In C-ImmSim, since $\bar{x} = x_0$ and $\sigma = 1$ for all cell populations, it is $E[x(t) \mid x_0] = \bar{x}$ and $Var[x(t) \mid x_0] = (\tau/2ln2)(1 - e^{-2tln2/\tau})$.

In this process, when the number of cells of type i is larger than the initial value $x_i(0)$, the death rate increases and $dx_i/dt < 0$; vice versa, if the number of cells is smaller than the initial value, it is the birth rate which increases and $dx_i/dt > 0$. Such a simplified behaviour fulfils the purpose of simulating both the increased/decreased production of cells, and the accelerated/ reduced rate of apoptosis in *in vivo* homeostasis.

Note that in the presence of an immune response the number of lymphocytes is always greater than the initial value. This means that $dx_i/dt < 0$ and more cells are deleted than in the absence of an antigen. On the other hand, the clones of cells that are actually growing are those specific for the antigen that eventually will become memory. The combination of these two facts causes the repertoire to shift toward specificities that recognise the antigen. Therefore, the definition of homeostasis by means of a mean reverting process helps in clarifying the Memory anti Naïve effect discussed in section 1.50.

During an immune response following antigen recognition, stimulated cells duplicate, i.e., in a time step corresponding to about 8 hours, a cell creates a copy of itself. The duplication process, as well as most of the events in the simulation, is stochastic, that is, a given probability is assigned to the event such that its actual realisation is decided by random numbers, which is equivalent to flipping a coin (to simulate an event with probability p, a random number between 0 and 1 is generated; if this number turns out to be lower than p the event takes place, otherwise it doesn't; this is called a Bernoulli trial). Thus, the probability for a stimulated B-cell to start the mitosis cycle is given by the following probability:

$$\Pr[B\ divides] = F_1(n(x, t))$$

where $n(x, t)$ is the total number of cells (of all types) in site x at time t and

$$F_1(n(x, t)) = exp\left(-\frac{n(x, t)^2}{\gamma^2(\Sigma_x n(x, 0))^2}\right).$$

The parameter γ determines a size-effect constraint on the clonal expansion and is chosen to allow a 4- to 15-fold increase of lymphocyte counts during acute infections [91]. After division, a B-cell matures into another B-cell or an antibody-producing plasma cell. This choice is made with probability 1/2. Besides that, a B-cell is committed to be a memory cell according to a process that is described in section 3.11.

Duplicating B-cells, helper and cytotoxic T lymphocytes create a copy of themselves with a probability that is computed as above, but with a substantial difference. The probability for a T-cell to divide is computed taking into account also, the inhibition or stimulation by cytokines,

$$\Pr[Th\ divides] = F_1(n(x, t)) \cdot F_2(v(x, t))$$

and also

$$\Pr[Tc\ divides] = F_1(n(x, t)) \cdot F_2(v(x, t))$$

where $v(x, t)$ denotes the concentration of the interleukin-2 in site x at time t, and

$$F_2(v(x, t)) = 1 - exp\left(-\frac{v(x, t)^2}{\eta^2}\right)$$

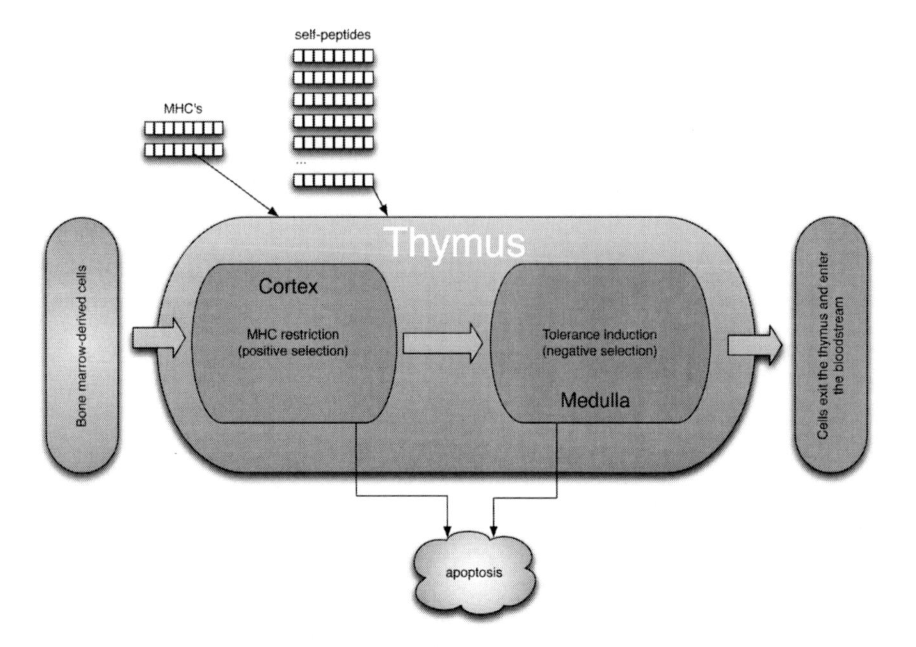

Figure 37. The two-layer filter realised by the thymus to eliminate auto-reactive T lymphocytes.

Colour image of this figure appears in the colour plate section at the end of the book.

represents the stimulation given by the local amount of cytokine IL-2 (also known as thymocyte growth factor). The parameter η represents the efficiency of the growth factor. Analogously, cytokine inhibition can be taken into account by simply using $F_2(c) - exp(-c)$ or another inverse relationship with the concentration of the inhibitor c.

3.8 The Selection of Cells In the Thymus

As already mentioned, the lymphoid cells undergo a process of maturation and education in the thymus prior to release into circulation. At variance with the thymus in IMMSIM which has three layers, one of which is optional, in C-ImmSim the selection process is modelled as a *two-layer filter*. The two layers' deeds are:

1. Positive selection of cells showing complementarity with the MHC part of MHC/peptide complexes;
2. Negative selection of cells showing complementarity with self-peptides complexed by MHC.

This process allows T-cells, as a population, to develop *self-tolerance* (in the negative selection) while eliminating useless cells (which fail positive selection). Self is defined by specifying as the input of the simulator, a set of binary strings, subset of all possible strings in the set $\{0, 2^{\mathrm{NBIT}} - 1\}$. The goal of the thymus selection is to filter out those cells that are able to match any one of the bit-strings defining the self.

While developing in the thymus gland, T-cells that match—even partially—the MHC (positive selection), learn to cooperate with cells expressing MHC molecules and proceed in the selection process. T-cells that react to the thymus' major histocompatibility complex, together with a self-peptide molecule, are eliminated (this is the negative selection). The result is that surviving T lymphocytes tolerate the body's cells and cooperate with them when needed. However, some T lymphocytes lose this ability to differentiate self from non-self, which results in the autoimmune diseases [91].

The match is between T-cell receptors and the molecular complexes formed by the peptide with the major histocompatibility complex molecule (either class-1 or 2). The binding is allowed in particular regions of the MHC/peptide (see Figure 39).

This process is regulated by a parameter called T_{eff} that represents the thymus efficiency in the selection: higher efficiency means better filtering, that is, less self-reactive cells in circulation. From the point of view of the calculation of the survival rate of the immature cells entering the thymus, the negative selection is treated as if the thymus were composed by T_{eff} sub-layers (each simulating the encounter with many self peptides specificity because of the crowded nature of the thymus). This means that the survival probability is the T_{eff} power of some elementary probability that depends on the MHCs and self-peptides available. Because of this, there is a cut-off value of the thymus efficiency T_{eff} that depends on the binary string length NBIT (around 2×10^3 for the 8-bit case), beyond which all self-reactive T-cells are destroyed in the thymus.

The process of selection engages immature thymocytes. The selection is performed versus self-peptides expressed in combination with both MHC class-1 and 2 molecules (see Figure 38). Thymocytes exit the thymus in a mature form, expressing either the CD4 or the CD8 co-receptor, hence identifying themselves as helper or cytotoxic T lymphocytes. In C-ImmSim, this is modelled by performing a different selection with either MHC class-1 molecules to select CD8 T lymphocytes (cytotoxic) or MHC class-2 molecules to select CD4 T lymphocytes (helper).

Each T lymphocyte, whose receptor is created at random in the set of the binary strings of length NBIT, is allowed to enter the circulation (i.e., to reach the secondary organ as a mature thymocyte) if and only if, (i) it is positively selected and, (ii) 'head' is obtained when flipping a coin with a probability (i.e., survival rate) computed in the negative selection phase.

A cell with receptor TCR is positively selected if it is met by at least one MHC molecule that has a match to the TCR greater or equal to NBIT/2 – m_0 (otherwise the T-cell is useless because it will never interact with the MHC/peptide complex notwithstanding the shape of the peptide). In other words, a T-cell with receptor TCR passes the negative selection with a probability that is given by

$$\Pr[\text{pass negative selection}] = \left[\prod_{j,k} (1 - p_1(j,k) \times q_1(j,k))(1 - p_2(j,k) \times q_2(j,k))\right]^{T_{eff}}$$

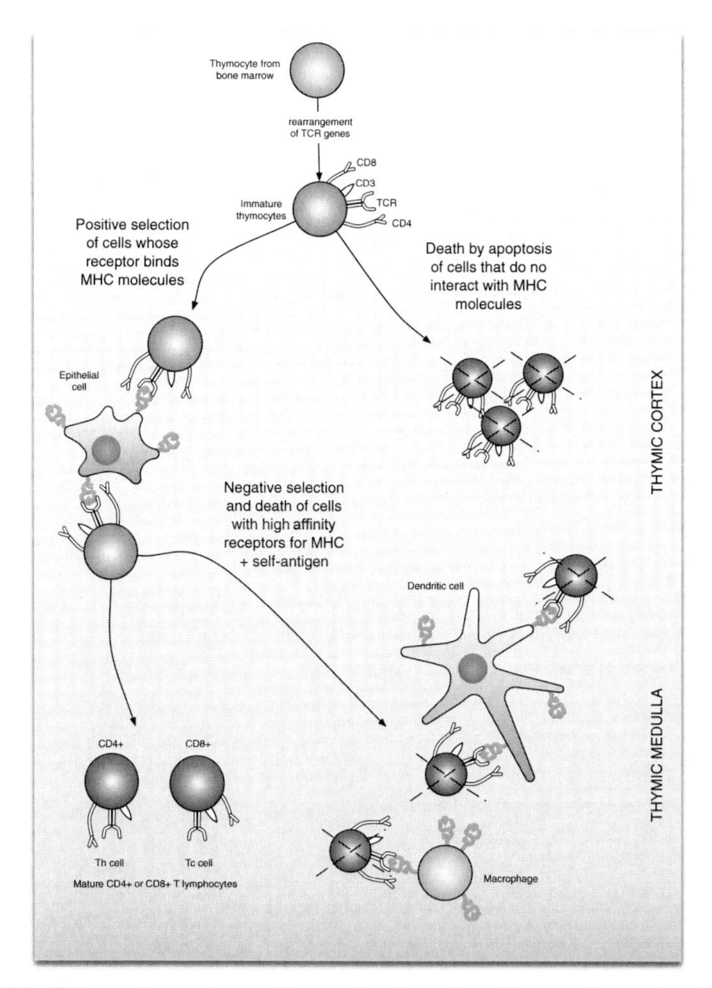

Figure 38. This cartoon shows the maturation of a thymocyte and subsequent involvement in the humoral and cellular immune response. In C-ImmSim, the maturation process is not explicitly modelled; just the activation from a resting state is necessary for a lymphocyte to enter into action.

Colour image of this figure appears in the colour plate section at the end of the book.

where for each molecule MHC_j and for each self-peptide pep_k

$$p_1(i, j) = \text{Affinity}(TCR, left(MHC_j) \odot right(pep_k))$$

$$q_1(i, j) = \text{MHCAffinity}(right(TCR), left(MHC_j))$$

$$p_2(i, j) = \text{Affinity}(TCR, left(MHC_j) \odot left(pep_k))$$

$$q_2(i, j) = \text{MHCAffinity}(right(TCR), right(MHC_j))$$

where $\odot$ is the bit-string concatenation operator, and the functions Affinity and MHCAffinity are the ones discussed in the previous sections. Going into details, $p_1(i, j)$ computes the probability that the TCR binds the binary string composed by the part of the MHC to the left and the part of the peptide to the right and $q_1(i, j)$ is the probability that the latter actually bind. The same, but for the part of the MHC to the left and the left side of the peptide, is given by the expressions for $p_2(i, j)$ and $q_2(i, j)$. This alternate matching comparison is depicted in Figure 42.

The elevation to the power of T_{eff} is made to simulate the cell passing through T_{eff} sub-layers. The result of having more layers is a more stringent selection of receptors. Since the process is stochastic, more layers reduce the probability of the selection of a cell. For example by taking $T_{eff} = 100$ (and having chosen two MHC-1 molecules), the procedure tests 1542 cells before selecting the 876 cells that go into circulation. Among these, just 12 are able to interact with the self; hence the fraction of self-reactive cells is 1.4%.

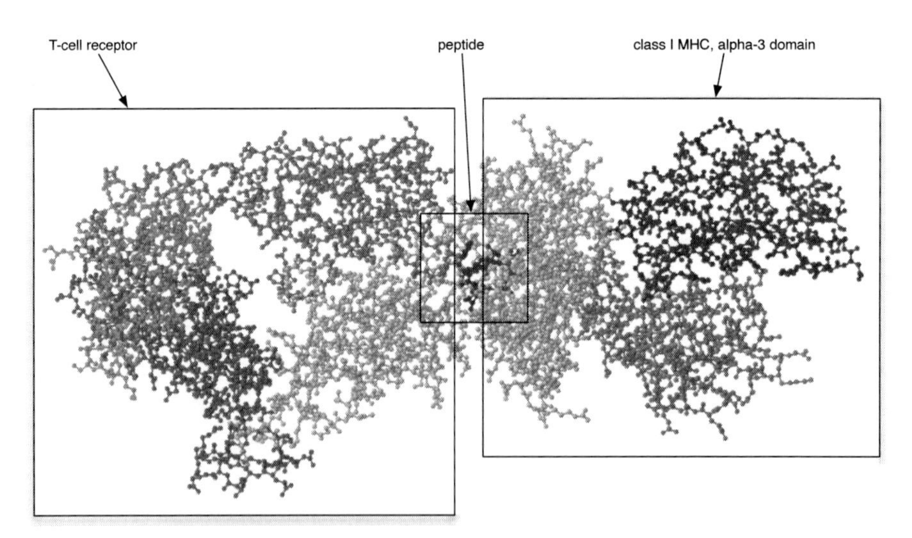

Figure 39. TCR binding is allowed in particular regions of the MHC/peptide. In the model approximation, receptors and peptides are binary strings. They stand for the effective binding regions only. Figure made with Jmol: an open-source Java viewer for chemical structures in 3D. http://www.Jmol.org.

Colour image of this figure appears in the colour plate section at the end of the book.

3.9 The Hayflick Limit

Wikipedia says that the Hayflick limit was discovered by Leonard Hayflick in 1965 when he observed that cells dividing in cell culture divided about fifty times before dying. Hayflick also observed that by approaching this limit, cells showed more signs of old age. The limit (which has been linked to the shortening of telomeres, a region of DNA at the end of chromosomes) varies from cell type to cell type, and more significantly from organism type to organism type.

A notable exception to the Hayflick limit in humans and other organisms is provided by stem cells as, by definition, they have not yet been fully differentiated, and therefore may continue to regenerate new cells for the entire lifespan of the organism, without limit. While the manifestations of the constant regenerative effects of stem cells is most easily seen in tissues which must continuously produce replacements for existing cells such as skin and blood cells, stem cells of one form or another are found in every tissue of the human body, even if only as dormant stem cells known as 'spore-like cells' [105].

Interestingly, cancer cells constitute the other main exception to the limits on cell division. From another viewpoint, it is believed that the Hayflick limit exists principally to help prevent cancer. If a cell becomes cancerous and the Hayflick limit is approaching, the cell will only be able to divide a certain number of times. Once it reaches this limiting number of divisions, the formed tumour will no longer be able to expand and the cells will die off. On the other hand if cancer grows it is because cells have found ways around the Hayflick limit. These are referred to as 'immortal'.

The Hayflick limit is implemented in C-ImmSim for all duplicating cells. This is a very simple rule. Just count the number of times a cell has entered the duplication cycle and limit it to fifty [106].

3.10 Cell Aging and Death

Aging and death of cells is ruled by the following simple differential equation on $x_i(t)$, the number of cells or molecules of class i at time t,

$$\frac{dx_t}{dt} = -k_i x_i$$

which has an exponential decay as solution $x_i(t) = exp(-k_i, t)$ with the constant k_i equal to $\log 2/\tau_i$ and τ_i the half-life, that is the time it takes for the exponential decay of to be reduced by half (i.e., $x_i(t + \tau_i) = x_i(t)/2$). In practice, at each time step, a number of entities of type i are eliminated. They are chosen randomly among the whole population but, for lymphocytes, the probability of elimination depends on their actual age a, that is,

$$P_{a,\tau_i}[\text{die}] = \frac{a^2}{K\tau_j^2 + a^2}$$

where K is an arbitrary parameter. Most of the values τ_i for cells or molecules are known from the literature [91, 92] and are reported in Table 6. The exceptions are the half-life of memory cells and plasma cells. Plasma cells are believed to live for few days although a different estimation has been made [107]. As regards memory, it is known that some memory cells live for years or even decades but it is very difficult to actually estimate this value [107]. Moreover, the immune memory of past infections certainly depends on the infectious agent encountered, as well as on the severity of the correlated affliction. For this reason, it sounds inappropriate to set any specific value for the half-life of memory cells. A more dynamic approach is necessary. This is what the following section is about.

3.11 A (Dynamic) Immune Memory

As far as the immunological memory is concerned, while researchers completely agree on the importance of some occurrences (increased number of specific B and T-cells available, persistence of activated cells), on other points, especially

Table 6. Half-life of cells and molecules (approximate values). Different types of immunoglobulins have different half-life. 'Antibody' refers to IgG. Values were taken from different literature sources [91, 92].

Cell/molecule	Half-life (days)
B lymphocyte	3
T helper lymphocyte	3
T cytotoxic lymphocyte	3
Macrophage	3
Epithelial cell	3
Plasma B-cell	3
Generic antigen	300
Immuno-complexes	30
Antibody IgM, IgG1, IgG2, IgE	10, 21, 20, 3
Natural killer cell	3
Danger signal	1
IL-2, IL-12, IFN-γ, IL-4, TGF-β, TNF-α IL-10, IL-6, IFN-β, IL-18, IL-23	3

those invoked to justify long-lasting memory (persistence of antigen, continuous stimulation, long-living memory cells), the consensus is far from complete.

The simulation of memory in IMMSIM is based on direct observations of adoptive memory in mice [108, 109] which revealed the following key phenomena: stochastic differentiation of long-lived memory cells during the primary response, facilitated help provided by the parallel helper and effector expansion at challenge, and di-phasic memory decay (see also sections 1.37–1.40). As a result, duplicating lymphocytes generate memory cells that have a longer half-life and memory cells all have the *same* value for τ_i.

C-ImmSim realises a different mechanism. It considers *memory* as the state of a cell that is acquired during *active participation* in successive immune responses. In particular, lymphocytes' memory is modelled by increasing the half-life by a certain amount every time that the cell successfully participates (i.e., interacts) in the antigen recognition. The bottom line is that useful cells survive longer than the useless ones, simply because they get a whole lot of stimulation during the immune response. Do recall what was said in section 1.7, that lymphocytes' specificity is rewarded by prolonging the life of those cells which bind the antigen. In practical terms, the parameter half-life, τ, of a lymphocyte (B, Th or Tc) is updated by a factor $\Delta\tau$ each time it *individually* interacts with another cell, thereby contributing to the recognition of an antigen and mounting an effective immune response against it. Hence, $\tau' = \tau + \Delta\tau$, where the increase is proportional to the affinity ($\Delta\tau \propto$ *affinity*) between the binding site of the lymphocyte (i.e., its receptor) and that of the interacting entity (e.g., a MHC/peptide complex on APCs). The overall result of this process is that few cells increase their half-life considerably and live longer than any other cell. Moreover, the expansion of the memory compartment is proportional to the magnitude of the infection and consequent duration of the immune response.

3.12 The Hyper-mutation of Antibodies

Hyper-mutation is a term used to indicate a set of complex phenomena whose result is a mutation in the portion of the DNA of the B lymphocyte coding for the *variable region* of the antibody [110] (see section 1.19).

In C-ImmSim, the mutation of each string representing the cell receptor in a duplicating B-cell is implemented as a stochastic process. The number of mutations follows a binomial distribution with parameters NBIT (the number of repetitions) and p_m (the probability of flipping a single bit). This schema assumes that the probability of single bit mutation events is *identical* for all bits of the same string. Note that this is a simplification, because in real life, certain segments of the receptor are certainly more susceptible to mutation than others. Moreover, a point mutation may influence the 'stability' of other portions so as to invalidate the assumption of independence. Nevertheless, this scheme allows interesting investigations about the effects of hyper-mutation on the overall immune response [7, 111] to be performed.

To show the effect of hyper-mutation on the maturation of affinity in C-ImmSim as well as in IMMSIM, one can use a "hole in the repertoire" by artificially preventing the bone marrow from generating high affinity to the antigen. As a consequence, hyper-mutation becomes the only mechanism for the immune system to enhance the affinity to the antigen. Mutations in B-cell receptors can either decrease or increase the affinity to the antigen. A less-frequent increase confers upon the cell an evolutionary advantage that eventually produces a shift of the overall population affinity toward a maturation of affinity.

3.13 Immune Activation

In theory, in absence of antigenic stimuli, the immune system remains 'still' (S1 in Figure 40). This state is *metastable* because any antigenic insult triggers an adaptive memory-generating immune response that reshapes the affinity landscape. The resulting immune state (S3 in the same figure) is more stable

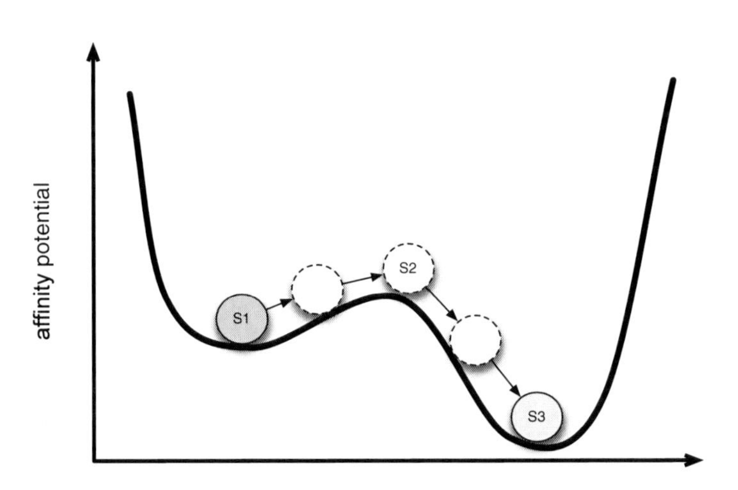

Figure 40. A metastable state of weaker recognition of the antigen (S1), in the initial adaptive immune response, corresponds to a local optima. It can shift through intermediate 'saddle' configurations (S2) to reach a stable state (S3) of stronger recognition corresponding to a better global affinity to the antigen.

Colour image of this figure appears in the colour plate section at the end of the book.

because the resulting system is more robust than before to the assault of the same or other cross-reactive invaders. To climb the hill of affinity potential, the system has to spend energy. This explains why the innate immunity tries, by any means, to avoid calling for the action of the adaptive immunity. Before this happens, the immune system is in a kind of resting state, waiting for activation signals.

3.13.1 Cell Activation by Danger

The activation process in C-ImmSim follows the danger model (or theory) of Polly Matzinger [12]. The idea behind this is that APCs respond to 'danger signals' specifically released from cells undergoing injury, stress or 'bad cell death' (as opposed to apoptosis, controlled cell death). The alarm signals released by these cells let the immune system know that there is a problem requiring an immune response.

In the model, this is translated into having macrophages becoming activated when they sense damage in the surrounding area. The source of damage is a necrotic cell or an *adjuvant* that is purposefully added to the injected molecules. The danger released by dying cells accounts for infected cells that are exploding because of their viral content or infected cells that are being killed by cytotoxic cells. Also, cancer cells that are killed by CTLs release a danger signal. The adjuvant itself is represented in C-ImmSim as a source of a danger signal. Therefore, an injection of adjuvant triggers a cellular response as if a cell had released it. The effect of danger on macrophages' activation is indirect, as natural killers will secrete IFN-γ in response to danger and IFN-γ will drive a state change in the macrophage from RESTING to ACTIVE. More precisely,

$$\text{If}(\textit{danger}) \text{ then } \textit{Nk} \text{ releases IFN-}\gamma$$

the probability for a resting macrophage being activated is

$$\Pr[\textit{MA activates}] = 1 - \nu \cdot exp(-(IFN_x / IFN_{eff})^2)$$

where IFN_x is the amount of danger signal in site x, IFN_{eff} is a parameter indicating the efficiency of the danger signal in activating the MA, and v is a parameter used to set the baseline macrophage activity. Conversely, the deactivation of an active MA is made with a fixed rate v'. To simplify things, if one takes $v - v'$, then v' can be used as a measure of the tendency of macrophages to remain in the RESTING state and $1 - v'$ is the fraction of macrophages that are active.

3.14 Anergy

Anergy is a form of cellular unresponsiveness following an antigen encounter. It is a tolerance mechanism in which the lymphocyte is functionally inactivated under certain conditions but remains alive for an extended period of time.

With regard to the anergic phenomena, a distinction between B-cell anergy and T-cell anergy must be made. While B-cell anergy is driven by a massive antigenic stimulation, T-cell anergy (both helper and cytotoxic T-cells) is determined by a lack of co-stimulation. This is achieved by putting the T cell in the anergic state if it has not yet taken an activation signal by macrophages or the dendritic cells.

3.14.1 B-cell Anergy by Overstimulation

B-cells enter the anergic state if over-stimulated by the antigen. This is computed at each time step and for each B-cell that is in the ACTIVE state, and it is extended to all antigens in the same lattice site (i.e., antigens that are potentially in contact with the cell). Providing details, indicating $A(e)$ the number of antigens with epitope e, the probability that a B-cell with receptor r goes to the ANERGIC state is

$$\Pr[B\ anergic] = \frac{1}{\Sigma A(e)} \sum A(e) \cdot \text{Affinity}(e, r)$$

where the summation spans over all antigens contained in the same lattice site of the B-cell. Once a cell becomes anergic it does nothing but sleep (i.e., no interactions, no duplications, etc.). An anergic B-cell can exit the anergic state with a fixed probability

$$\Pr[B\ exists\ anergy] = 1 - exp(-\log(2)/h_{½})$$

where $h_{½}$ s the halving time of the population of anergic cells (e.g., $h_{½} = 3$ time steps equivalent to about a day means that the number of anergic cells halves in a day).

3.14.2 Helper T-cell Anergy by Lack of Co-stimulation

Induction of anergy in helper T lymphocytes is thought to occur during activation in the absence of adequate co-stimulation. Helper T-cells go ANERGIC if they are stimulated by B-cells that are exposing the MHC-2/peptide complex while they are RESTING. In fact, T helpers first need a specific activation signal by the macrophages. This first signal wakes them up from the resting state. Once activated, they can interact with B-cells and initiate an immune response. If the B-cell signal arrives *before* a macrophage activates them, then the helper T-cell goes into the ANERGIC state with the probability

$$\Pr[Th\ anergic] = \text{Affinity}(r, u^{II})$$

where r is the TCR and u^{II} is the MHC-2/peptide complex. As for the B-cells, T helper cells exit anergy with probability given in the equation above.

3.14.3 Cytotoxic T-cell Anergy

CTLs enter the ANERGIC state by following a similar rule. They go ANERGIC if they receive a signal while in the RESTING state. In this case, it is not the B-cell that stimulates them but any cell that is exposing an antigen peptide on class-1 MHC molecule (e.g., infected cells, malignant cells, etc.). So, again, a cytotoxic T-cell goes anergic with probability

$$\Pr[Tc\ anergic] = \text{Affinity}(r, u^{I})$$

where r is the TCR and u^{I} is the MHC-1/peptide complex exposed on the cell surface. To switch back from RESTING to ACTIVE, the T-cell has to be activated (first signal) by macrophages, that is, it needs to interact with MAs that are presenting the peptide on the MHC-1 molecule. The probability of this event follows the general rule for computing the affinity, that is, the probability is equal to the affinity of the TCR with the MHC-1/ peptide molecule on the surface of the presenting macrophage.

3.14.4 Bystander Effect

Stimulated T helper cells produce cytokines. All the interleukins produced, among other functions, control the strength of the bystander effect. In C-ImmSim, this is the activation of cells without direct cell-cell contact. In other words, a helper T-cell or a B-cell can be stimulated to divide if the concentration of stimulated T helper cell in its proximity (i.e., the lattice site) is high (i.e., above a given threshold $\varphi \in [0, 1]$). The probability for a B or a Th to enter the mitotic cycle without direct cell-to-cell stimulation is computed as

$$Pr[duplicate] = 1 - (1 - \varphi)^{T_x}$$

where T_x is the number of Th-cells in site that are in the DUPLICATION state. The only extra requirement for a B-cell to be stimulated to duplicate by the bystander effect is that it has to be in the PRESENTING-2 state, that is, presenting the antigen peptide with the MHC class-2 molecule.

3.15 Interactions among Entities

The core of the whole simulation is represented by the set of rules coding the interactions among entities. In the model, these are executed in a *randomised* order at each time step to avoid any consequent systematic behaviour [112].

The interactions among entities are described in terms of state transitions as already discussed. They can be divided into two categories: *external interactions*, which happen among cells and molecules having the same position on the lattice and *internal (to the cell) interactions*, which account for MHC Ag-peptide digestion *inside* the presenting (i.e., APCs) or infected (e.g., Ep) cells. They are listed in Table 7 together with other important procedures of the model, implementing the features or processes described in the preceding sections.

Two interacting entities can bind in a *specific* or *non-specific* manner. In other words, if the interaction is subjected to the binding of a membrane receptor (as in the case of B-Ag and successive B-T interaction in Figure 41), the interaction is specific, and it is unspecific otherwise (think, for example, of

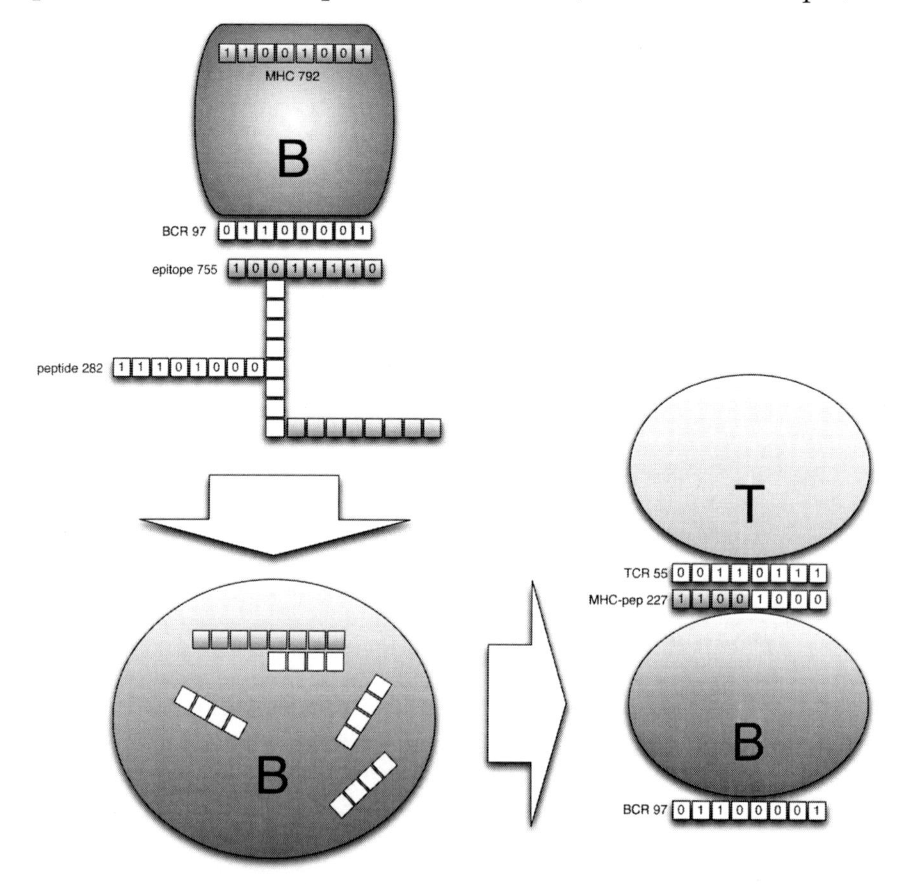

Figure 41. A B-cell binds the antigen. After the antigen digestion and presentation together with the MHC molecule on its surface, it interacts with an active Th cell upon matching with the T-cell receptor. This is the 'pivotal' cooperation step discussed in section 1.16.

Colour image of this figure appears in the colour plate section at the end of the book.

Table 7. Interactions among cells, or cells and molecules; antigen digestion and presentation on MHC class-2 for presenting cells and class-1 for infected cells; other procedures of the model including, for example, the duplication of cells, the generation of new cells, etc.

Cells' and molecules' interactions
B phagocytosis of antigen
MA phagocytosis of antigen
Dc phagocytosis of antigen
B presentation to Th
MA presentation to Th
Dc presentation to Th
Formation of immune-complexes
MA phagocytosis of IC
Infection of endothelial cells
Cytotoxicity action of Tc
Inhibition of Th and Tc by Treg
B phagocytosis of Ab (idiotype network only)
B phagocytosis of IC (idiotype network only)
Cells' activations and anergy
Activation of MA
Anergy of B-cells
Anergy of Th cells
Priming of Th cells
Anergy of Tc cells
Activation of Nk by danger
Activation of Tc cells
APCs' antigen digestion and presentation
B exogenous pathway
MA exogenous pathway
Ep endogenous pathway
Other procedures
Haematopoiesis
Clone division
Plasma secretion of immunoglobulins
Diffusion of molecules
Cell movements
Antigen replication

the phagocytosis of antigens by macrophages). By evaluating the bit-string match, and assigning a probability for the action to take place, the recognition of matching conformations is performed (section 3.5).

Each interaction procedure is executed on each lattice point in serial order. Since the interaction takes place among entities on the same lattice site, the scanning order of the lattice sites does not alter the dynamics of the computation. In contrast, since the state of each cellular entity (and the total number of molecules in a lattice point) determines the next macro state of the system as a whole, the dynamics strongly depends on the order in which the interaction procedures are executed. For instance, if one first executes the interaction between macrophage and virus (i.e., phagocytosis) and then the interaction between the virus and the macrophage as target cell (i.e., infection), then the virus is penalised with respect to the opposite case.

The interactions follow a *Greedy* paradigm [113], meaning that when comparing two sets of cells, the first successful matching is followed by the removal of the interacting couple from the two sets.

The following pseudo code implements an interaction between the entities type T1 and T2 (e.g., B and Th lymphocytes). First observe that the function uses the local lists of cells on the lattice point indexed by x, i.e., entities interact locally. The greedy algorithm requires a complete scanning of the list of lymphocytes in the search for the first matching. Eligible cells must be in one of the allowed states (e.g., to keep up with the B-Th example, the B has to be in PRESENTING-2 and the Th ACTIVE). Then, the affinity between the binding sites of the two entities, a_1 and a_2 (e.g., T-cell receptor against the MHC-2/peptide complex on the B-cell), indicated by Affinity(a_1, a_2), is computed as in section 3.5, and the interaction occurs probabilistically (i.e., a Bernoulli trial with probability p). Upon success a number of events take place, counting entity state-changes, counter increments/decrements, release of cytokines, etc. All these are incorporated in the function interact (a_1, a_2). The greedy algorithm then removes a_2 from the list L_2 and proceeds with the next a_1 in list L_1.

algorithm Interaction(T1 T2)
be T1 and T2 **the** entity types
do for each site x
    L_1 = list of entities in x with type T_1 in the eligible state s_1
    L_2 = list of entities in x with type T_2 in the eligible state s_2
    do for each a_1 in L_1
        do for each a_2 in L_2
            p = Affinity(a_1, a_2)
            if COIN(p) **then**
                interact(a_1, a_2)
                $L_2 = L_2 - a_2$
                next a_1
            fi
        od
    od
od

In a more abstract way, to describe what is called the 'external' interactions, the following syntax is adopted:

ENTITIES : < *involved entities* >

SPECIFIC : < *Yes* | *No* >

MATCH : < *involved molecules* >

CONDITION : < *allowed states for* < *involved entities* >>

ACTION : < *new state for* < *involved entities* >>

where *involved entities* are the *two* interacting entities (no 'three-in-a-bed' interactions are implemented); the field *SPECIFIC* is *Yes* if the interaction probability depends on the matching degree between the *involved molecules* of the two entities and is *No* otherwise (*unspecific* bind). The *involved molecules* are the binding sites of the *involved entities* such as the membrane receptors of lymphocytes and epitopes of the antigen.

The conditions *allowed states for < involved entities >* are expressed in terms of *first order predicates* by means of the logical Boolean operators (dyadic connectives) *AND*, *OR*, *NOT* and of other *unary operators* (unary connective) that check each possible state of the entities looking at the entities' *flag-bytes*. The state transitions are registered by flipping the corresponding flags in the flag-byte. In case the entities have more than one receptor (or generic binding site) like the antigen that might be represented by two or more epitopes, the interaction is allowed only if *at least one* of their binding-site matches. In particular, the actions are undertaken as soon as one match is successful, again following the *greedy* paradigm.

Each of these procedures examines the data structures of the two involved entities, looking for simultaneous true values of the predicates expressed in *CONDITION*. For each *specific* interaction, the Hamming distance between the involved molecules specified by the tag *MATCH* is computed. After that, the probability of a successful interaction is obtained by looking at the affinity function. At this time, a random number between 0 and 1 is generated. If the random number is less than the interaction probability, then the actions specified in the tag *ACTION* are undertaken. These actions are usually composed of very few assignments so they do not constitute a major overhead compared to the scanning of the two lists. For every *unspecific* binding (i.e., *SPECIFIC : No*), the probability to bind does not depend on any match and the field *MATCH* is empty.

As a practical example, one can consider the interaction/recognition between B-cells and the antigens (Figure 41). In the formalism introduced above, this interaction is described as follows:

ENTITIES : *B*, *Ag*

SPECIFIC : *Yes*

MATCH : *BCR*, *epitope*(*s*)

CONDITION : $B = ACTIVE\ OR\ B = PRESENTING\text{-}2$

ACTION : $B \rightarrow INTERNALISED\ AND\ Ag \rightarrow deleted$

In each site of the lattice, the procedure may find a couple of elements in the two sets of B-cells and antigens for which the conditions are fulfilled. In this case the conditions are: the B-cell has to be active or presenting the epitope and the match is between its receptor and the antigen epitope. If more epitopes define the antigen, then all of them are tested; the set of epitopes is scanned in random order to avoid bias. Given the BCR and the Ag's epitope, the match and the affinity are computed in terms of probability to bind, utilising the affinity function as defined in section 3.5. Then a *coin* with that probability is flipped and the decision of what to do is taken. If no bind takes place, the cell is confronted by other antigens in the lists. If the cell, instead, binds the antigen then the following actions are taken: the B-cell changes state, the antigen is deleted and a pointer to the bound antigen structure is stored in the local memory structure of the B-cell. The procedure then continues with the next B-cell until all B-cells are scanned.

A list of all the interactions (external and internal) is reported in Table 7.

3.16 Antigen Digestion and Presentation

The syntax for the description of the cell-internal events is very similar to the previous case. The only difference is that one of the involved entities is always the infecting/engulfed antigen. This is, from the computational viewpoint, a major advantage, since just one data structure must be scanned to select which cells will be processed. The following is an example, not much different from the other cases.

Internal digestion of the antigen peptides by antigen-presenting cells like the B lymphocytes or macrophages is done according to the following logic:

ENTITY : *B*

SPECIFIC : *Yes*

MATCH : *MHC-2(s), peptide(s)*

CONDITION : $B \equiv INTERNALISED$

ACTION : $B \rightarrow PRESENTING\text{-}2$

Analogously, the endogenous antigen is processed and presented on the class-1 MHC molecule (this is the cytosolic or endogenous pathway) instead of the class-2 MHC molecule as in this case of exogenous antigen processing (endocytic or exogenous pathway). The various procedures are reported in Table 8.

Table 8. Antigen digestion and presentation on MHC class-2 for antigen-presenting cells and on MHC class-1 for infected cells.

Digestion and presentation
B exogenous pathway
MA exogenous pathway
Dc exogenous pathway
Ep endogenous pathway

Note that in the binary representation, the peptides are arranged in two substrings (see Figure 42): one binds to the MHC and the other is shown in the MHC–peptide complex. In contrast, the MHC molecule offers just one binding site for the peptide that is conventionally chosen to be the right half of the binary string. Therefore, when the right half is bound to one half of the peptide (either left or right), the one that is shown in the MHC/peptide complex is the left part.

3.17 Cell Motion and Diffusion of Molecules

There are several aspects that must be taken into account when describing cell motion. First of all, cells do not have all the same mobility. Several studies looked at real cell dynamics using two-

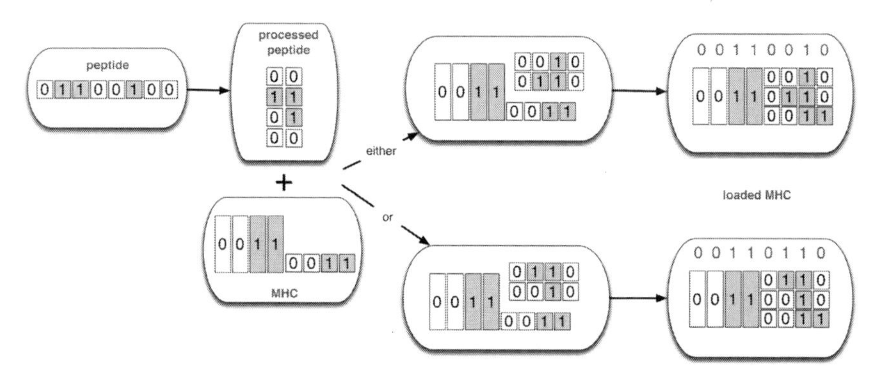

Figure 42. In the binary representation of the peptides, they are arranged in two substrings: one binds to the MHC and the other is shown in the MHC–peptide complex. In contrast, the MHC molecule offers just one binding site for the peptide, that is conventionally chosen to be the right half of the binary string. So, when the right half is bound to one half of the peptide (either left or right), the one that is shown in the MHC/peptide complex is the left part (adapted from [96]).

Colour image of this figure appears in the colour plate section at the end of the book.

photon microscopy and showed marked differences between different lymphocytes. T helper cells appear to be the fastest, with an average velocity of 11 $\mu m/min$, followed by B-cells with 6 $\mu m/min$ [114] and Dc with a velocity of 3 $\mu m/min$ [115, 116].

In absence of chemotaxis, entities move freely in all directions, provided the number of cells at one point does not exceed the available volume. Since the physical space is an input parameter, and because the number of cells can grow in number during an immune response, the lattice may become filled to a limit concentration. The probability of diffusing from x' to a randomly chosen neighbour x for each cell and for each time step is

$$\Pr[x' \rightarrow x] = 1 - o_x$$

and o_x is computed as follows:

$$o_x = \begin{cases} N_x/\widetilde{M}, & N_x/\widetilde{M} < 1 \\ 1, & N_x/\widetilde{M} \geq 1 \end{cases}$$

where

$$\widetilde{M} = \frac{10^9 \cdot n \; [\mu\text{m}^3][\text{cells}]}{L^2 \cdot 523.6 \; [\mu\text{m}^3]}$$

is the maximum number of cells in the volume represented by each lattice point and N_x is the total number of cells (of all kinds) in the destination lattice point x. $\widetilde{M}$ is computed assuming a radius of 5 μm equal for all cells meaning a volume of about 520 cubic micrometres. Therefore, if n cubic millimetres are simulated, that is $n \cdot 10^9$ cubic micrometres per cell, then a lattice point can host at most $\widetilde{M}$ cells (supposed to be incompressible). Molecules are 3–4 orders of magnitude smaller, so they are not considered in this approximated calculation.

The second aspect has to do with the overall direction of motion traced by lymphocytes and the role played by chemokine gradients. Imaging experiments have revealed a typical stop-and-go motility [115]: T-cells seem to randomly move for few minutes, then pause, feel chemokine gradients and then suddenly turn in a new direction. Lymphocyte traffic thus looks random-like on long time scales but chemotaxis still plays a central role in setting the overall direction of motion and the correct timing of lymphocyte interactions.

The description of chemotaxis phenomena is rather difficult because of the difference of time scales involved. Cells display a diffusion coefficient between 10 and 70 μm^2 per *min* whereas the typical coefficient for molecule diffusion ranges from 600 to 6000 μm^2 per *min* [117]. It has been estimated that a single cell can effectively communicate up to a distance of about 250 μm in 10–30 minutes [117].

When chemotaxis is taken into account, the diffusion process becomes biased towards the direction of maximum chemotactic gradient. Once the chemokine distribution has been computed, cells can be moved according to the chemotactic signal. Each cell can be sensitive to a specific chemokine and therefore should move towards its increasing gradient. This is accomplished by multiplying (and rescaling) the occupation effect included in the previous equation for o_x by the chemokine gradient ∇c

$$\Pr[x' \rightarrow x] = (1 - o_x) \cdot \nabla c_x$$

where ∇c_x indicates the (normalised) difference in the chemokine concentration moving from x' to the neighbour site x. It is particularly important to be model this process when chemotaxis plays a crucial role, as in the case of cell motion within lymph nodes [118]. In this case, each cellular agent is equipped with a specific label that says whether the receptor for that chemokine is expressed on its surface or not. For instance, a T-cell becomes sensitive to chemokines released by MA once activated by the first 'effective' encounter with a dendritic cell. As that happens, its receptor for these chemokines is shown on its membrane, and the receptor for T-area chemokines is deactivated. T-cell motion, therefore, is chemokine/receptor dependent.

Molecules like interleukins and chemokines rapidly diffuse in all directions on the mesh representing the simulated volume, whereas their target cells may need several minutes to cover a few microns. Therefore, the range of influence of a molecular signal like interleukins can be very large, and lymphocyte motion can be affected by chemokine signals released far away from their position. Moreover, during its slow displacement, a cell feels a chemokine gradient that is not constant but changes continuously in time and space. To address this issue, a very simple choice would be to reduce the time scale unit to the smallest time scale involved. Both cells and molecules could be described individually and the two dynamics followed with high precision. However, this approach would greatly slow down the computation, both because of the reduced temporal step and the large number of molecules involved. The more effective approach, followed in C-ImmSim, is to introduce a multi-scale description of both the chemokine diffusion and the cells' chemotactic motion. While cells are treated with the usual stochastic discrete agent-based approach, chemokine diffusion is modelled using partial differential equations. The actual diffusion dynamics thus takes into account the proper diffusion coefficients and the correct time scales of motion and signalling (for molecules). Moreover, the chemokine concentration is better characterised from a spatial point of view, leading to an improved description of the chemotactic gradient felt by cells.

In C-ImmSim, it is possible to pick a time step in a range between 120 and 480 minutes and cell diffusion processes are automatically rescaled. Once the time step is selected, the lattice unit is set equal to the distance that a T helper cell (i.e., the fastest cell entity in the model) can cover in one time step (Figure 43). The average distance covered by a particle moving in random motion, during a time interval dt is given by $< dx > = \sqrt{6Ddt}$, where D is the *diffusion coefficient* of the particle. Since the mesh unit corresponds to the average distance Dx covered by a T helper cell, it is $D = D_{Th}$. The motion of B-cells, plasma cells, macrophages and dendritic cells is relative to T-cell diffusion, e.g., B-cells take approximately eight time steps to cover the same distance dx.

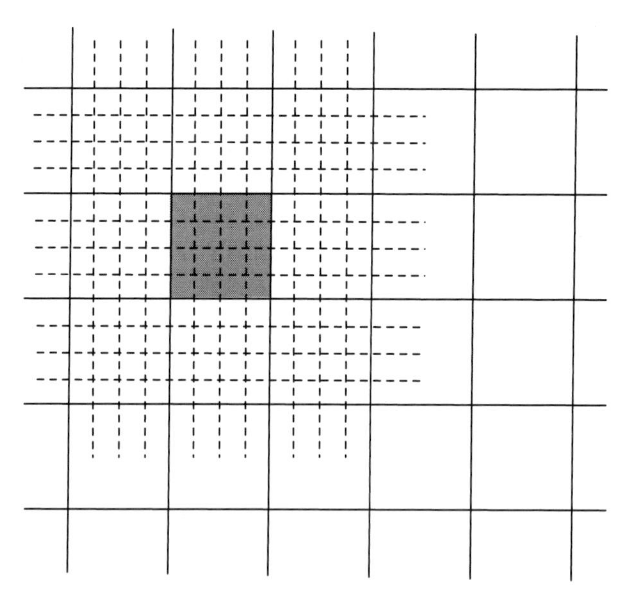

Figure 43. Integration mesh. The original mesh (thick lines) is refined and a smaller is used (dashed lines). In this way, the spatial integration of the partial differential equation (PDE) for chemokine diffusion is more accurate.

Chemokine diffusion is described by a partial differential equation, meaning that the chemokine concentration is a continuous quantity which changes in a deterministic way. The equation describing the diffusion is a simple parabolic equation (similar to the heat equation) plus a degradation term that takes into account the finite half-life of the molecules:

$$\frac{\partial c}{\partial t} = D \cdot \Delta c - \lambda c$$

where D is the diffusion coefficient and λ is the half-life [117, 119].

3.18 Main Procedures

The whole simulation is built on a loop that iterates all main procedures. Their execution order is randomly determined at each time step to avoid bias due to narrow synchronisms. Moreover, as already mentioned, since each of these procedures updates the state of the cells in a *serial* way (i.e., cells are organised in first-in first-out FIFO structures), the lists of cells are continuously shuffled to avoid giving advantage to cells that are in a predefined order in the lists.

```
Read input parameters
Initialise data structures
do repeat (until simulation ends)
begin
  1. Inject antigen if scheduled
  2. Cell generation in the bone marrow
  3. T-cell maturation in the thymus
  4. Cell to cell/molecule interactions
  5. Antigen digestion and processing
  6. Antibody secretion by plasma cells
  7. Clone division
  8. Antigen proliferation
  9. Aging and death by apoptosis
 10. Diffusion of entities
end
```

After reading the user-defined parameters, the code allocates the data structures and variables and reads all necessary input from storage. In this phase, the initial conditions are set, the entities are generated and the lattice is populated. Once all is set, the main loop is executed. The procedures in the do-loop are iterated for a number of steps or until another termination condition is reached. Generically speaking, other termination conditions can represent the situation in which a biologically-meaningful range is exceeded so that the virtual experiment has to come to an end.

The procedure at line 1 consists of checking whether an injection is scheduled. The injection consists of the antigen (plus an adjuvant if planned). In particular, it consists of the type of antigen (virus, bacteria, inactive compound, allergen, etc.), its number of particles, its composition (the exact form of its epitopes and peptides) and other possible parameters (e.g., the viral replication rate). Another piece of information contained in the injection record is whether the antigenic content should be uniformly distributed on the lattice grid or 'placed' at a single lattice point. This difference simply translates in a delayed immune response and possibly in a different final outcome since, for example, a bacterium has the time to proliferate beyond the system capacity.

The procedure at line 2 consists of calculating the number of new cells at each time step, their creation, initialisation and positioning on the lattice. The calculation of the number of new cells is shaped to guarantee homeostasis as described in 3.7.

Newly generated T lymphocytes have to go through the thymus selection before reaction the periphery as mature cells (procedure at line 3).

Cells residing on the same lattice site are allowed to interact (line 4). Whether they actually do it depends on a number of conditions which will be specified later. Note that this is the real core of the model as it puts into *words* that immunological knowledge which is the matter of active investigation (Who's doing what? Who's interacting with whom?).

The procedure at line 5 entails APCs' digestion of previously phagocytosed (exogenous) antigen (this is the *exocytic pathway*), binding to class-2 MHC molecule and exposure of the MHC/peptide complex on the membrane cell surface for presentation to T helper lymphocytes. Likewise, endocytic antigens (e.g., viral peptides) are processed (*endocytic pathway*), bound to MHC class-1 molecules and shown on the membrane to cytotoxic T lymphocytes.

If plasma cells are present, they secrete large amount of antibodies against the recognised antigenic peptide (line 6).

Cells that have been stimulated to proliferate do so (line 7). A cell becomes two by replication of the data structure in memory. Quite simple. Subtle differences are taken into account on a case-by-case basis.

Antigens also duplicate (line 8). Bacteria duplicate, viruses replicate, tumours grow and so on. All kinds of bad things happen.

Time goes by (line 9) and cells, like anything else, age. (Which is not so bad, after all.) A counter is incremented and some of them eventually die with a probability computed on the basis of a formula that takes account of the time counter and the cell relative parameter half-life, as described in section 3.10.

Finally, in the procedure at line 10, all cells and molecules move from one lattice site to a randomly chosen neighbour site. Each entity moves independently and with a cell-specific diffusion coefficient. The entire process resembles a Brownian motion of particles. Besides that, cells sensitive to specific *chemotactic signals* stochastically follow a preferential direction given by the chemokine gradient.

3.19 Notes on the Expressed Repertoire

How many cells are created during a simulation? Both cells that initialise the simulation and those that are created at runtime from the bone marrow have a receptor that is chosen at random. The probability of generating a receptor that has a perfect match with the antigen is

$$p = \Pr[m = \text{NBIT}] = \frac{1}{2^{\text{NBIT}}}$$

therefore, populating the space with $B(0)$ randomly generated B-cells, one gets on average

$$p \cdot B(0) = \frac{B(0)}{2^{\text{NBIT}}}$$

cells with a receptor that matches perfectly the epitope of the antigen. Moreover, during a simulation according to the haematopoiesis process, on average,

$$\sim B(0)\,\frac{\text{In}(2)}{\tau_B}$$

B-cells with a random receptor for each time step are generated, where τ_B is the average half-life of B naïve cells (i.e., cells that have not interacted with the antigen yet). For example, if $B(0) = 10^{\circ}$ and $\tau_B = 10$, one gets that at each step (each equivalent to ~8 hours), about 70 naïve B-cells are generated. It follows that over t time steps, on an average,

$$t \cdot B(0)\,\frac{\text{In}(2)}{\tau_B} \cdot p$$

cells whose receptors have a perfect match with the antigen are created. These are to be added to those created at $t = 0$ to get the total number of perfect match cells that were (and still are) available to the immune system to recognise and react to the antigen

$$p \cdot B(0)\left(1 + t\,\frac{\text{In}(2)}{\tau_B}\right)$$

For example, given NBIT = 8, $B(0) = 10^3$, $\tau_B = 10$ and $t = 200$, an average $10^3/2^8\ (1 + 200 \cdot \text{In}(2)/10) \sim 58$ cells with affinity NBIT are created within 200 time steps. Let's make another example, NBIT = 12, $B(0) = 10^3$, $\tau_B = 10$, $t = 500$; then, on an average, $10^3/2^{12}\ (1 + 500 \cdot (\text{In}(2))/10) \sim 87$ cells with a perfect match are created in 500 time steps. That's a small number! Consider however that the probability that not even a single perfect match string is created in t time steps is

$$(1-p)^{B(0)\left(1 + t\,\frac{\text{In}(2)}{\tau_B}\right)}$$

which in this example means that the probability to fall short of a 12-bit perfect match in 200 time steps is $1.7 \cdot 10^{-4}$. Note that when NBIT is high, p is very small, hence the number of cells generated with a perfect match is small. In that case, the maturation of affinity plays a very important role since it produces high-affinity clones by mutation of low-affinity ones which are much more numerous (see 1.19).

3.20 Scaling the System Size

Since the chance of a successful encounter between the antigen and a matching lymphocyte, in the space of the simulation, depends on the number of cells and the lattice dimension, these conditions, together with the bit-string length NBIT and the minimum match m_0, are all important parameters in determining the time of the onset of the primary immune response. These parameters can be set by comparing the delay in the emergence of the immune response with that of a *qualitative reference case*. This reference scenario consists of a few observations obtained when performing the classical immunisation experiment:

i. *A primary response developing in 2 or 3 days*
ii. *Peaking at day 4 or 5*
iii. *Clearing the antigen in not more than a week*
iv. *A secondary immune response faster then the first.*

Having set the bit-string length NBIT and the minimum match m_0, the number of B-cell receptors recognising the antigen's epitope(s) depends not only on the concentration of cells and antigens on the lattice (as in the Michaelis–Menten kinetics) but also on the initial repertoire.

A reference point is given by a volume of one microliter, NBIT = 8, $m_0 = 7$, the size of a square lattice $L = 16$ and $B(0) = 10^3$. This case shows a dynamics that is qualitatively in agreement with major stylised facts of the vaccination experiment, i.e., antigen clearance time, magnitude of the immune response, time for relaxation, persistence of memory, and so on. In practice, the size L is computed by solving the equation

$$\frac{B(0)}{L^d} \cdot K(\text{NBIT}, m_0) = C$$

where

$$K(\text{NBIT}, m_0) = \frac{1}{2^{\text{NBIT}}} \sum_{j=m_0}^{\text{NBIT}} \binom{\text{NBIT}}{j}$$

gives the fraction of cells that have a receptor with a match above threshold m_0, i.e., that can interact with the antigen, and the constant C

$$C = \frac{10^3}{16^2} \cdot K(8, 7)$$

is the concentration of useful cells per unit of volume in the reference case.

Therefore by imposing the same concentration, the value of the lattice size, L, can be calculated as a function of NBIT, m_0 and the initial population of B-cells $B(0)$, that is,

$$L = \left[K(\text{NBIT}, m_0) \frac{B(0)}{C} \right]^{1/d}$$

where, by the leukocytes formula $B(0)$ is typically taken equal to 260 *cells*/μl which refers to the normal concentration of B lymphocytes in the blood of a human [120].

3.21 A Few Words on the Definition of Molecular Quanta

Antibodies as well as other molecules represented in the model are handled in 'quanta' of concentration, that is, the algorithm treats them as multiples of a predefined unit amount. These unit amounts differ among molecule types and have been chosen to match with experimental data as in [102, 121]. For example, they are some picograms for antibodies and some femtograms for cytokines. As can be readily surmised, there is some arbitrariness in this definition but that is unavoidable, given the variability of experimental data. For example, with regard to antibody production, bibliographic sources indicate that a single plasma cell is able to secrete $2-8 \cdot 10^3$ immunoglobulins per second; that is equivalent to about $6-24 \cdot 10^7$ in eight hours. Now, the mass of a single antibody is about $2.5 \cdot 10^{-7} pg$ and a time step of the simulation is about 8 hours, therefore a simulated plasma cell secretes on average, 40 picograms in a single time step; hence a quantum of antibodies is of the order of a picogram of mass.

3.22 The Choice of the MHC Molecules

The choice of the MHC molecules is important, as the binary strings representing the haplotype determine the likelihood that the immune system recognises the antigen injected into the system. To be more precise, the antigen immunogenicity is determined by the sum of three different things: (i) the choice of the self-peptide strings shaping the space of all possible antigens' peptides in what is self and what is non-self (remember the section dedicated to the thymus selection 3.8); (ii) the efficiency of the thymus selection procedure, that is, a high efficiency means a clear-cut separation between self and non-self while low efficiency means that auto-reactive cells also slip out of the thymus; (iii) the MHC molecules defining the individual haplotype. This means, that before starting a simulation by choosing the antigen peptides and the MHC molecules and, at the same time, a certain set of self-peptides, one needs to check if the antigen peptides are not too similar to the self-peptides in which case no immune response should be expected. Also, if the MHC's match to the antigen peptide is too small (i.e., weak binding), then the immune response will be weak and slow. In contrast, an antigen equipped with peptides that have a high affinity to the MHC binary strings, and that are at the same time very different from self-peptide strings, will be highly immunogenic, that is, it will elicit a strong immune response.

3.23 The Parameters

Like most complex simulators, this model has many parameters. Most of them are somehow 'hard-wired' into the code so they should not be called parameters because they are, in practice, constant. However, sometimes, it is convenient to be able to change them to test extreme conditions or unusual hypotheses. For example, one may want to experiment with an immune response in which the ability of APCs to phagocytose an antigen decreases with the age of the virtual patient.

In general, the parameters of this model can be sorted out in three categories: (i) parameters whose value is known from standard immunology literature [91, 93, 98]; (ii) parameters intrinsic to the process of model development that have no correspondent counterpart in reality (these are somehow arbitrary and set to plausible values after performing a series of tests); and (iii) parameters corresponding to the initial and boundary conditions which determine the problem under investigation.

For the arbitrary parameters, it is important to mention that this model has been used in different studies, so that most of these parameters have already been set and do not need to be selected again [1, 7, 82, 94-97, 112, 122, 123]. This is the advantage of using the same model to study the immune system in different disease conditions since the immune system is 'one', and there is no reason to question a mechanism that has been shown to characterise well a certain phenomenon that has been the matter of investigation in the past. Think for example, of the per-bit mutation probability in the hypermutation phenomenon or of the thymus efficiency or of the macrophage Ag-engulfing efficiency in a viral infection. In the absence of experimental data evidence, there is no reason to question those values. Indeed, often, 'characterising well' a certain phenomenon requires the validation of the model against a set of data coming from the clinic or from the laboratory. Setting parameters in such complex models requires some confidence and trust: if you trusted your validation yesterday, why should you not trust it today?

3.24 What to Monitor?

During the runs, a number of variables that fully define the on-going infection and the immune response can be monitored, for example: the rate of killing of epithelial and antigen-presenting cells, the antibody *titers*, the proviral and free virus copies, the number and the specificity of activated and anergic Tc, the fraction of APC cells that are presenting the antigen, the number of infected cells, the number and specificity of Th, the number and specificity of B-cells, the number of resting B and MA cells and so on.

A lot of other data can be derived through a built-in logging mechanism that keeps track of all events during the simulation. While extremely heavy computationally, this option allows one to carefully inspect subtle details of the dynamics which are otherwise easily overlooked.

4

Benchmarks, aka Qualitative Model Validation

This chapter contains some comments on the problem of validating the computational model. This is one of the hardest parts of the development of a simulator in general. Since *in vivo* or *in vitro* experiments are money and time consuming, many parameters are quite difficult, if not impossible, to be determined. For this reason, some parameters are to be considered arbitrary while others can be chosen according to what one can find in a normal textbook on immunology or in research papers. This has just been discussed above.

One of the advantages of ABMs lies in the fact that the interactions of agents are derived from fundamental occurrences in biological processes, like the binding of receptors, and as such, they are more intuitive and easier to understand. Additionally, the instructions that describe the interactions are taken from published literature and translated into programming language [62]. One direct consequence of this approach is that the resulting dynamics can be read in analogy with experimental reality in a straightforward way. Cell counts are cell counts. The number of infected cells is just so. Occasionally, a cell's fate can be traced individually and milieus that are nowhere to be found in mean-field approximations can be discovered. This allows following long-ranging effects, either in time or in space, by tracing small local perturbations.

Due to the inherent stochasticity of the C-ImmSim model, parameter exploration and model calibration is often performed on a trial and error basis. This process involves the systematic execution of the simulation, varying the values of key parameters, thus comparing (and selecting) multiple sets of results. Moreover, averages are computed to define mean behaviours in the stochastic dynamics as well as its standard deviation. Note that the variability in the results has a different meaning than the biological variability estimated from available 'real data', since the sources of the latter are orders of magnitude more numerous [62].

4.1 Primary and Secondary Response

In this example, an antigenic 'substance' or vaccine is initially injected to elicit a primary response. At a later time, the injection is repeated to test the memory effect in the secondary response. This experiment follows the prime/boost scheme of a typical vaccination and evidences how the immune system develops memory according to the clonal selection theory.

On day three, an arbitrary dose of an innocuous but immunogenic antigen is injected. One can ignore, for the moment, the specific 'shape' of the injected molecule, that is, the strings composing it. What is important is to note how this injection elicits recognition by antigen-presenting cells and humoral response by the immune system (Figure 44). The whole antigen amount is cleared in less than a week during the primary immune response by the antibodies created but also, although to a minor extent, by antigen-presenting cells such as macrophages and dendritic cells (Figure 47). These cells phagocytose the antigens, digest them, and later present them on class-2 MHC molecules. Antibodies are still present when a second injection is performed. In fact, some of the vaccine inoculated is lost because antibodies coat the antigenic molecules

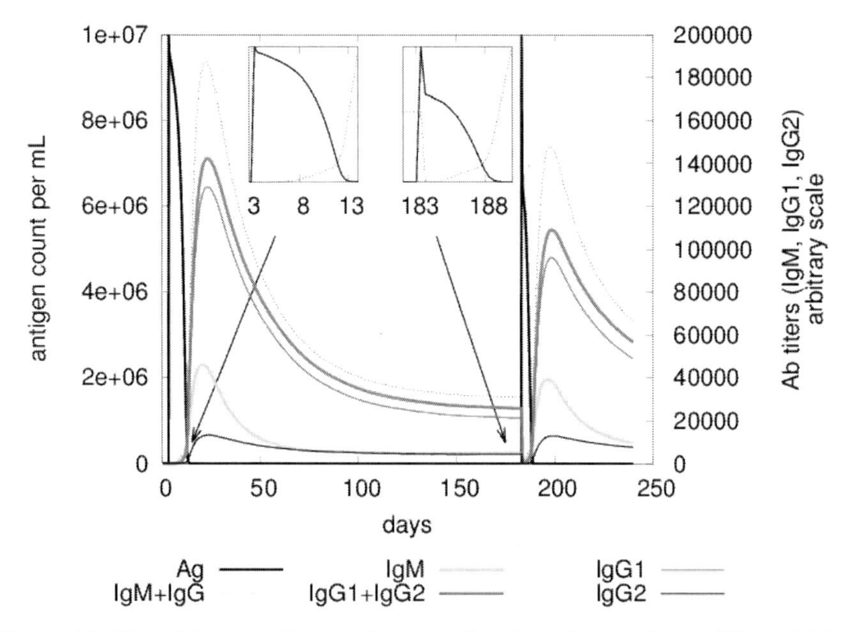

Figure 44. This plot shows the number of antigenic molecules in a millilitre of blood versus time in a vaccination experiment. Two injections are administered at different time steps to observe a faster secondary immune response due to the memory effect elicited from the priming injection. This plot shows, also, the total number of specific antibodies produced (on the y2 axis).

Colour image of this figure appears in the colour plate section at the end of the book.

actually competing with APCs. Despite this fact, the secondary response is about twice as fast and much stronger than the first. This is due to the presence of specific memory lymphocytes that are left from the first recognition (Figure 45).

Specific antibodies are secreted by plasma cells (left panel of Figure 48) produced by cloning B-cells shown in Figure 45. The panel on the right shows B-cell counts with respect to the internal state. It is instructive to note, in correspondence to the injection of the antigen, an increase of both B-cells that are presenting the antigens' peptides combined with class-2 MHC molecules on their surface and, with a little delay, cells that are duplicating. Duplication is allowed when presenting cells find a cognate receptor on T helper cells (Figure 46) that stimulates them to divide. Also, in Figure 45, note that both isotypes of

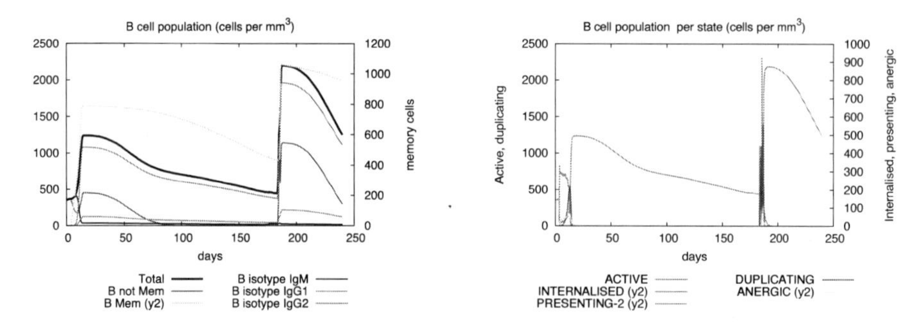

Figure 45. On the left is the number of B-cells in a cubic millimetre of lymph node tissue or (for convenience) in a micro litre of peripheral blood, together with the relative number of memory cells (on the y2 axis). On the right the same number is plotted, but for each possible internal state the B-cells can take.

Colour image of this figure appears in the colour plate section at the end of the book.

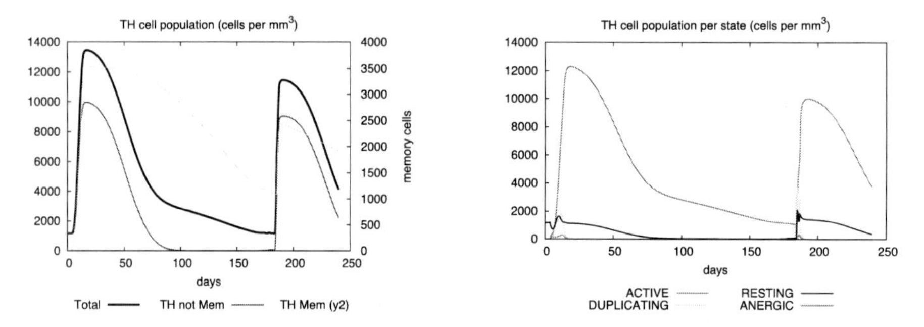

Figure 46. T helper lymphocytes during a vaccination experiment.

Colour image of this figure appears in the colour plate section at the end of the book.

B-cells, IgM and IgG, are shown. IgM is quickly exceeded by IgG. In the same figure can be spotted the emergence of memory cells whose half-life has been increased by the repeated contact with the antigen. After the prime injection and upon antigen's clearance, the stimulation ceases and the formed clone shrinks, leaving a small number of memory cells ready to be stimulated anew by the challenge injection about six months later. The right panel of Figure 48 shows the concentration of interleukins. Note that during the prime injection there is a higher peak of TGF-β compared to other cytokines (IFN-γ, IL-12, etc.) due to the regulatory action of Tregs. This is interesting because it illustrates how T-regulation intervenes during the prime injection. The massive dose of antigen injected requires a regulation of the immune system that, if left uncontrolled, could cause collateral effects. During the boost injection, this is not happening because the antigen dosage is largely dealt with by circulating antibodies and therefore there is no need to downregulate the immune response. During the boost response, the higher peak of IL-4 is within stochastic ranges, and therefore it is not informative of a polarisation toward the Th2 phenotype of helper T-cells.

In Figure 46, the same kind of information is reported for the T helper cells. Of note: a small fraction of T-cells became anergic because of an 'excess' of stimulation. This also happens to B-cells in Figure 45.

Figure 47 shows the macrophages (left panel) and the dendritic cells (right panel). As antigen-processing cells, these can be found in the state INTERNALISED preceding PRESENTING-2 and following the phagocytosis of the antigen. INTERNALISED is an intermediate state that is quickly changed (consider that a time step is 6–8 hours); this is the reason why the plots do not show any cell in this state.

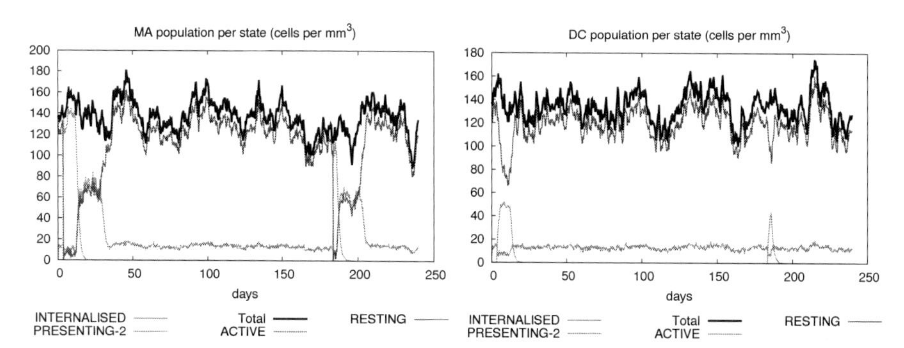

Figure 47. As antigen-processing cells, macrophages (MA, left panel) and dendritic cells (DC, right panel) can take many internal states like the intermediate state, INTERNALISED, preceding the state PRESENTING-2 and following the phagocytosis of the antigen. The INTERNALISED is an intermediate state that is quickly changed in a time step equivalent to 6–8 hours; this is the reason why the plot does not show any cell in this state.

Colour image of this figure appears in the colour plate section at the end of the book.

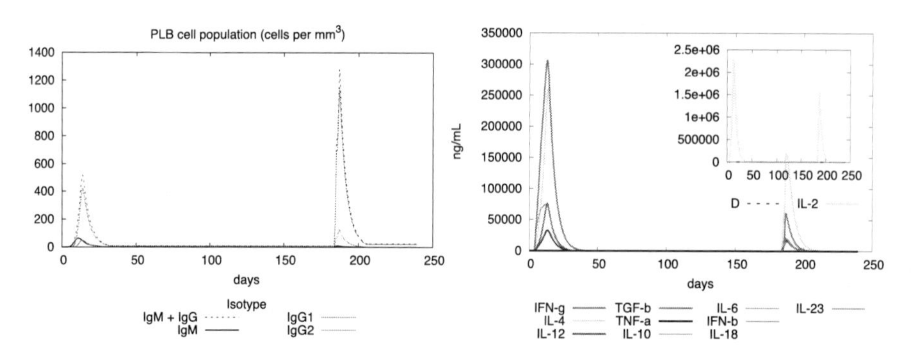

Figure 48. Plasma B-cells (PLB) peak after antigen stimulation. The predominant isotype is the IgG (left panel). In the right panel, the plot shows the concentration of various interleukins, each carrying a different message for the activated immune system.

Colour image of this figure appears in the colour plate section at the end of the book.

4.2 Exhaustion

The dose of the injected antigen plays a non-trivial role in determining isotype switches, Th polarisation, and so on. Moreover, while low dosages are not able to elicit a long-lasting memory, a high dose can make the system (i.e., cell population compartment) hit some constraints. This results in a kind of systemic saturation with consequential delay in clearing the antigen. This feature has been investigated by performing a study in which increasing dosages of antigen have been injected to the system in a single shot, and the total antibodies measured at population peak. The result is shown in Figure 49.

What is interesting to observe in this experiment is that the humoral immune response results tend to be limited because of constraints in the volume (i.e., by a *crowding effect*), because cell anergy originates from overstimulation or because the Hayflick limit is hit. This observation can be reasonably extended to the cytotoxic response as well. Viral immunologists' interest has recently focused on the competition for space between responding T clones, and the phenomenon of 'attrition' has been the object of a revealing interdisciplinary study [29]. About the modelling of attrition, see section 1.48.

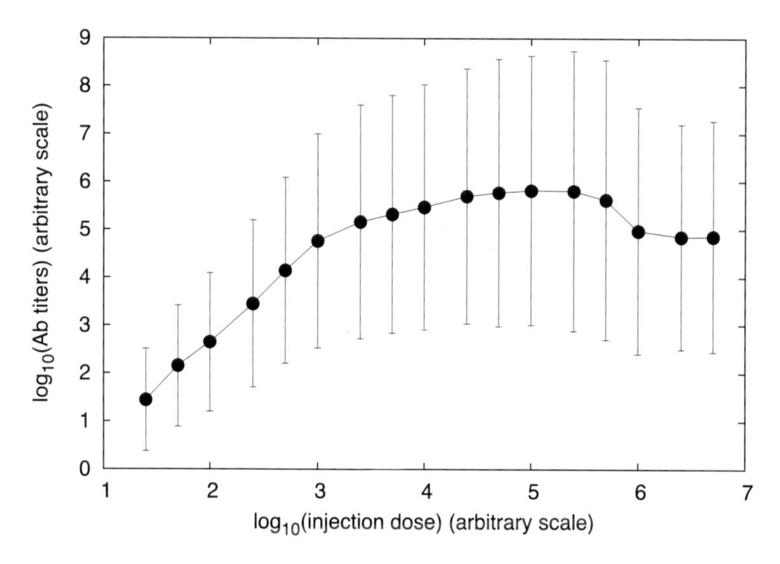

Figure 49. Peak values for antibody titers responding to a priming injection of a vaccine. The system saturates for volume constraints (overcrowding effect) and, later on, for very large amounts of injected doses; it even reduces the response due to anergy induced by overstimulation.

4.3 Bacterial Infection

To mimic a bacterium, one can simply allow the injected antigen to create copies of itself. Thus, a generic bacterium is an antigen characterised by its replication rate R. Its dynamics can be described with the Fisher's equation

$$\frac{\partial B(x,t)}{\partial t} - R \cdot B(x,t)\left(1 - \frac{B(x,t)}{B_\infty}\right) + D\,\frac{\partial^2 B(x,t)}{\partial x^2}$$

where $B(x, t)_x$ is the number of bacteria at time t at position x, D is the diffusion coefficient, and B_∞ is the carrying capacity of the system. Neglecting the diffusion term and the carrying capacity, one gets the solution $B(t) = B_0 \cdot e^{R \cdot t}$, which is a very simple exponential law of growth without self-limiting factor. In this experiment, a small quantity of bacteria is injected on day three. Since a replication rate R to the detriment of the bacterium has been purposefully chosen, the immune system succeeds in stopping the bacterial population growth within twelve days from infection.

The following plots show essentially the same results as in the previous vaccination experiment. The differences from the previous dynamics are evident and do not need more explanation than is provided in the figure legends.

While the reader is left with the task of interpreting the plots in Figures 50 through 54, and observing the few differences (mainly in magnitude of the response) from the previous experiment, a different unrevealed aspect of the simulation is now shown.

Figure 55 displays some snapshots of the spatial evolution of a bacterium in a simulation different than those just discussed. The figure shows the bacterial concentration $B(t, x)$ and the specific antibodies $Ab(t, x)$ produced by plasma cells, side by side. In the initial phase of the immune response, a relatively small number of bacteria is randomly distributed all over the space. Antibodies appear in spots, indicating local recognition by APCs and stimulation by T helper cells. Differentiation of

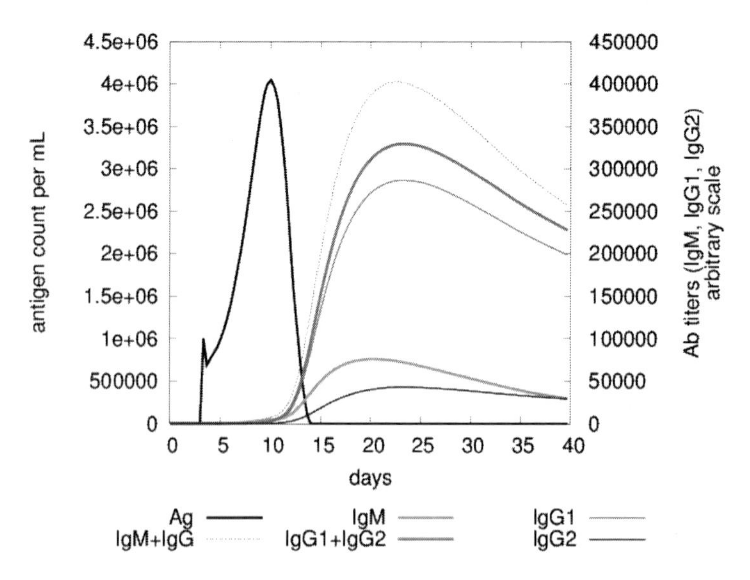

Figure 50. Bacteria replicate freely until an effective immune response is mounted about one week after the infection. The control is done at the limit. Another few days of exponential growth and the population of bacteria would have been uncatchable.

Colour image of this figure appears in the colour plate section at the end of the book.

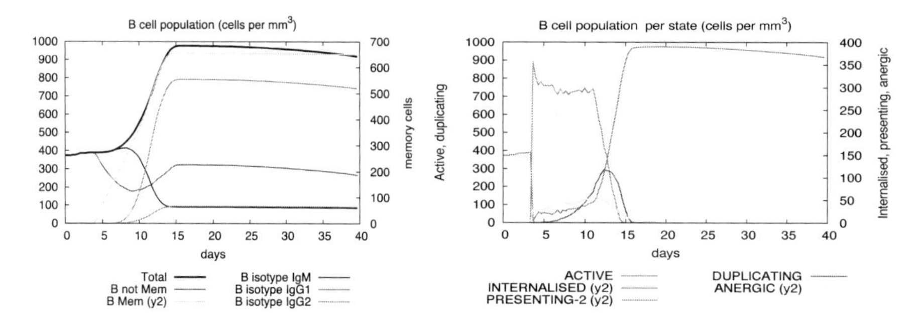

Figure 51. Left plot shows the virgin and memory B and the plot on the right shows the same population grouped by state.

Colour image of this figure appears in the colour plate section at the end of the book.

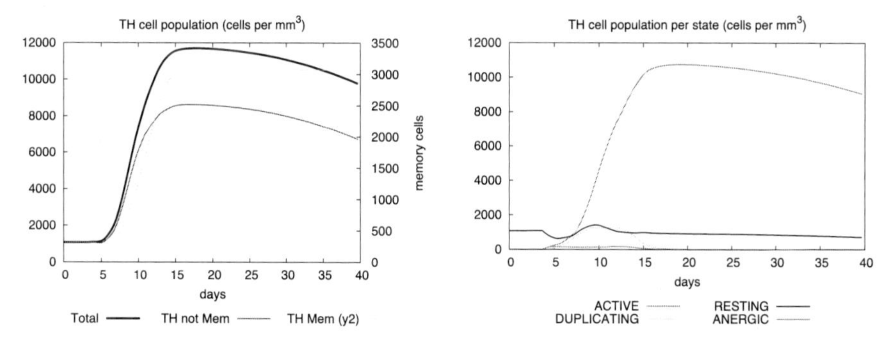

Figure 52. Helper T-cell dynamics during a bacterial infection. The plot on the left shows the virgin and memory T-cells whereas the plot on the right shows the same population by state.

Colour image of this figure appears in the colour plate section at the end of the book.

B-cells to plasma cells, which secrete large amounts of antibodies, follows. While bacteria proliferate, the production of antibodies increases. At a certain point, most of the bacteria are cleared but, interestingly, there are two areas in which the bacteria survive and proliferate. Eventually, this simulation will see bacteria grow beyond immune control (not shown).

This example shows pretty clearly how inhomogeneity and stochasticity together play a very important role in explaining dynamics otherwise unattainable with models lacking these features. These sorts of spatial effects play a very important

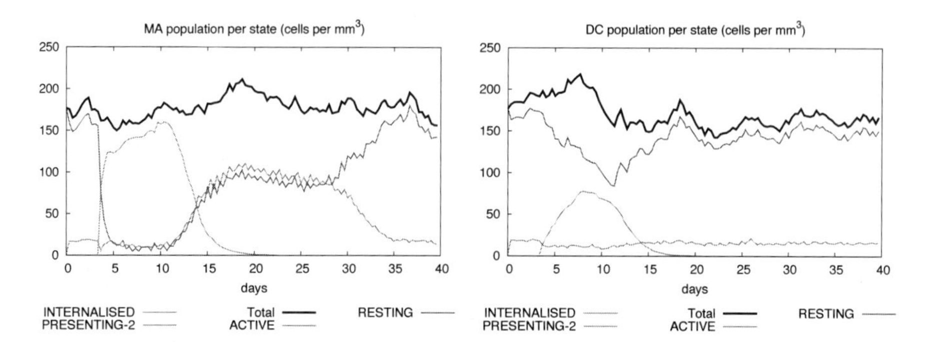

Figure 53. These plots show the dynamics of macrophages (MA, left) and dendritic cells (DC, right) during a bacterial infection. Dendritic cells are shown to do a better job of presenting the antigen. Both macrophages and dendritic cells present the captured antigen to helper T-cells to mount the immune response. Macrophages need to be activated before this work can begin.

Colour image of this figure appears in the colour plate section at the end of the book.

role in the pathogen dynamics, *in vivo* as well as *in silico*. For example, the human immunodeficiency virus (HIV) is able to escape immune control by accumulating in reservoirs or sanctuaries like the brain.

4.4 Viral Infection

C-ImmSim is equipped with the humoral response that was seen at work in the vaccination and in the bacterial infection of the previous examples, and with the cellular immune response to pathogens that infect host cells. This is a test of the ability of the model to deal with a virus. The responses spoil the replication machinery of the invaders and block their proliferation.

A minimal representation of the virus can be obtained by the following points, each specifying a peculiarity of the virus life cycle: (i) identify a target cell for the virus to infect; (ii) specify the infection rate of the virus to the target cell, that is, the probability that the virus gets inside the cell upon contact with its membrane; (iii) provide the rules for the dynamics of infected cells if the infection changes their normal behaviour; (iv) in case of a retrovirus, set its transcription rate; (v) set the replication

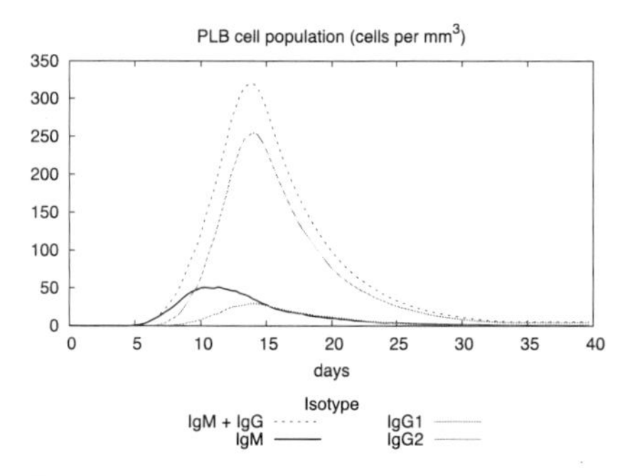

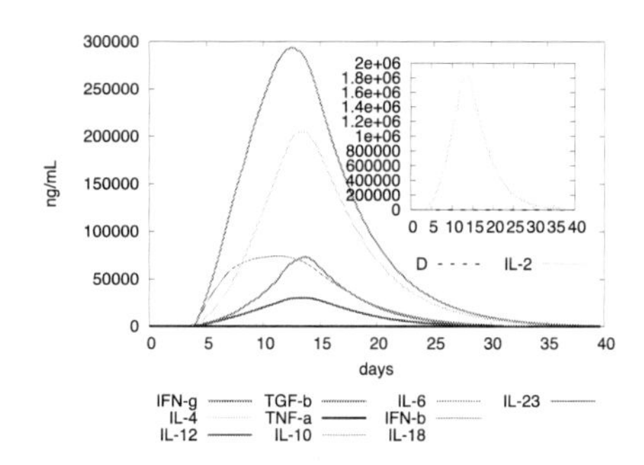

Figure 54. Plasma B cells (PLB, left) and cytokines (right) during an immune response to a fast-replicating bacterium. IgG1-producing PLB cells surpass IgM- and IgG2-producing cells. On the cytokines pattern, the presence of TGF-β and IL-4 reveals a Th2 response, but because pro-Th1 cytokines IFN-γ and IL-12 are also present, the response is not at all unbalanced.

Colour image of this figure appears in the colour plate section at the end of the book.

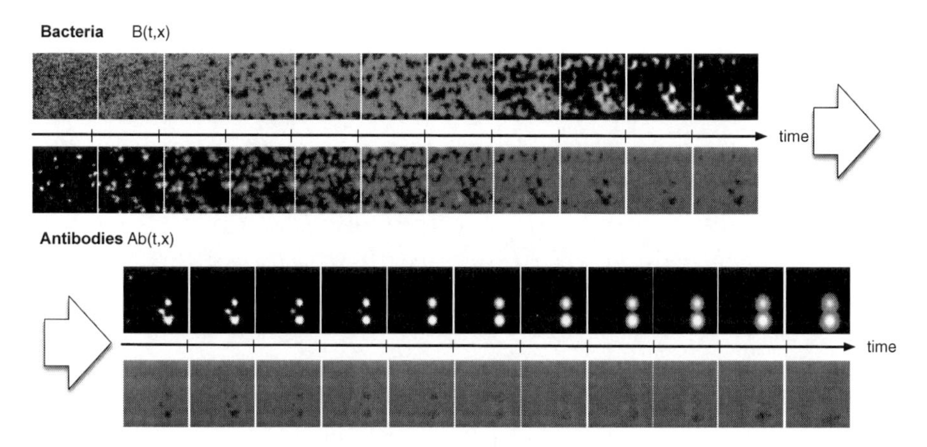

Figure 55. Spatial evolution of a bacterial infection. Twenty-two shots of the bacterial concentration, $B(t, x)$, and antibody concentration, $Ab(t, x)$, on a two-dimensional squared lattice (1000 × 1000). Dark/black areas indicate low concentrations. It is interesting to observe the formation of two spots in which bacteria survive and proliferate, notwithstanding a high concentration of antibodies.

Colour image of this figure appears in the colour plate section at the end of the book.

rate of the virus; (vi) assign a mutation rate to the virus; (vii) set a budding rate of newly assembled viruses from the cell membrane (how many fully functional viral particles?), or the bursting threshold, that is, the threshold value for the number

Figure 56. Illustrations of the coronavirus (left) responsible for the SARS, the *West Nile* virus responsible for the West Nile Fever (centre) and the HIV-1 virus responsible for AIDS (right). The fine structure of the viruses is not implemented in C-ImmSim: only their most important features like replication, infectivity and mutation are represented, together with details about their life cycles.

Colour image of this figure appears in the colour plate section at the end of the book.

of newly assembled viruses inside infected cells, above which the cell bursts and releases all viral particles in the environment.

What follows describes an experiment in which epithelial cells provide the target and the virus does not mutate.[5] In this experiment, the activation rate (i.e., the probability of activation) is equal to one, which is to say that once inside a target cell, the virus promptly starts replication. This kind of virus makes the target cell burst when a critical number of viral particles is reached.

The panel on the left of Figure 57 shows the free virions (what is called *soluble* virus because it is found in the blood), and the total number of virus inside infected cells (*proviral* virus). In this case, according to the specific values of the parameters, the near totality of viral particles is found outside infected cells. As can be readily deduced, this is the result of a high replication rate and a low cell-bursting threshold, which shorten the time a viral particle stays inside the infected cell. It is fun to play around with these parameters

[5] A technical note: The fact that the virus does not evolve by mutation is a non-trival simplification from the computational point of view, because the data structures of the algorithm can be simplified so as to deal with a single antigen entity; for a virus like the HIV-1, which mutates at a high rate, things are a 'bit' more complicated.

and observe the resulting dynamics. On the other hand, one can systematically explore the various outcomes as described in the work [124].

The right panel of Figure 57 shows the detailed dynamics of epithelial cells. Since they constitute the target of the virus, once they are infected and start presenting the viral peptides in the context of class-1 MHC molecule, they become the target also of cytotoxic cells (Figure 60). This is visible by the large fraction (almost the totality) of cells in the state PRESENTING-1. Once the virus is cleared at about day 20, the epithelial cell population is restored.

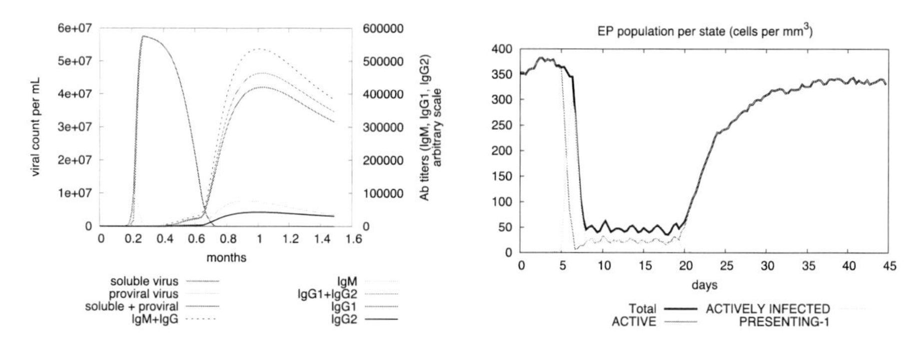

Figure 57. The virus infection triggers the production of immunoglobulins, mainly of the IgG1 type (left panel). In this plot, *soluble* means free virus while *proviral* indicates viral count inside infected cells. In the right panel, the detailed dynamics of epithelial cells (EP) as targets of the virus is shown; after a primary immune cytotoxic response (see Figure 60) which largely decreases their population, epithelial cells are kept at a low count by a continuous virus killing by blasting and by CTLs. This is visible by the large fraction (almost the totality) of cells in the state ACTIVELY INFECTED. On the other hand, a small fraction of the infected cells are able to present the virus peptides with the MHC-1 molecule on their surface.

Colour image of this figure appears in the colour plate section at the end of the book.

The virus stimulates a humoral response as well. High concentration of antibodies as against the viral epitopes can be seen in Figure 57. B-cells grow in number during a viral infection (Figure 58). In the initial acute phase, some become anergic because of the high concentration of viral particles.

Eventually, B-cells duplicate and differentiate to plasma B cells (PLB), producing antibodies. The other plots show the antigen presentation function of macrophages and dendritic cells in Figure 61, the enabling role of T helpers in Figure 59 and the already mentioned cellular immunity by CTLs in Figure 60.

In this case, the virus was easily cleared by the humoral/ cellular immunity, but in other cases it can lead the immune system to a metastable state of chronic activation. This happens, for example, if the transcription rate is small so that the virus

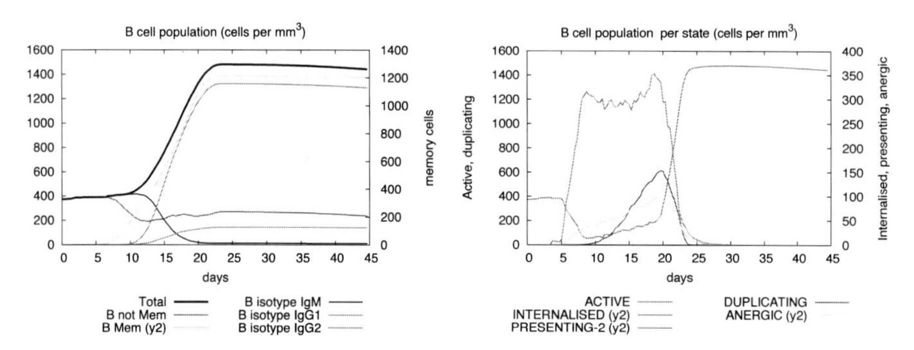

Figure 58. B lymphocytes grow in number during a viral infection. In the initial acute phase, some B-cells become anergic because of the high concentration of viral particles. Eventually, B-cells duplicate and differentiate to plasma B cells (PLB), producing antibodies.

Colour image of this figure appears in the colour plate section at the end of the book.

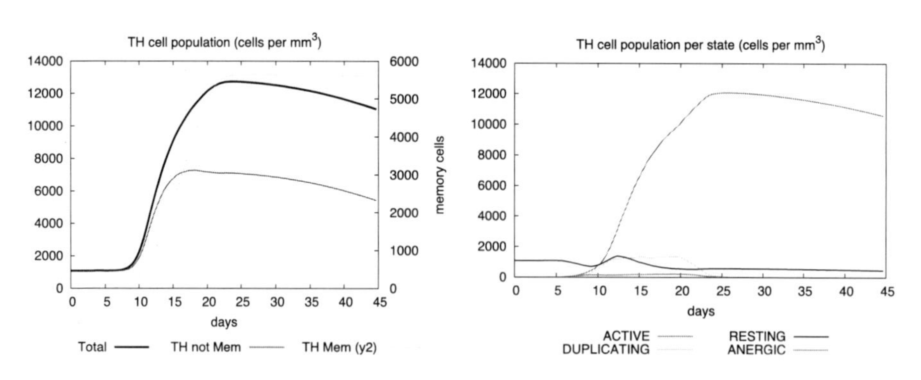

Figure 59. These are the Th cells during a viral infection. A large number of active T helper cells is deployed to eliminate the threat.

Colour image of this figure appears in the colour plate section at the end of the book.

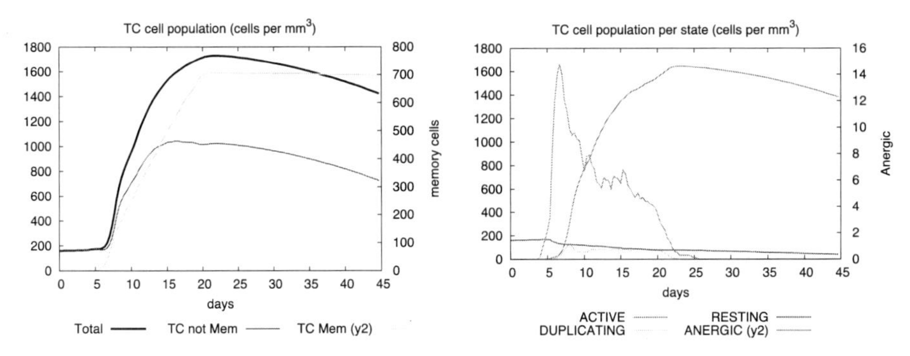

Figure 60. Cytotoxic T-cells responding to a viral infection. During the acute phase of the infection, some Tc become anergic by overstimulation. However, these anergic Tc are very few, and the majority of cytotoxic cells are actively killing infected epithelial cells.

Colour image of this figure appears in the colour plate section at the end of the book.

Figure 61. During the viral infection, macrophages and dendritic cells do their job by capturing soluble virus particles and presenting these antigens to helper T-cells.

Colour image of this figure appears in the colour plate section at the end of the book.

remains latent in cell reservoirs, or for more complicated definitions of the virus, like the HIV-1, which is able to mutate (section 5.2).

The sequence of pictures in Figure 62 shows the growing concentration of the virus in the simulated space by taking snapshots of the two-dimensional lattice at time intervals. Each pixel of the figures corresponds to a lattice site: black means lower virus concentration, green means high virus concentration. The

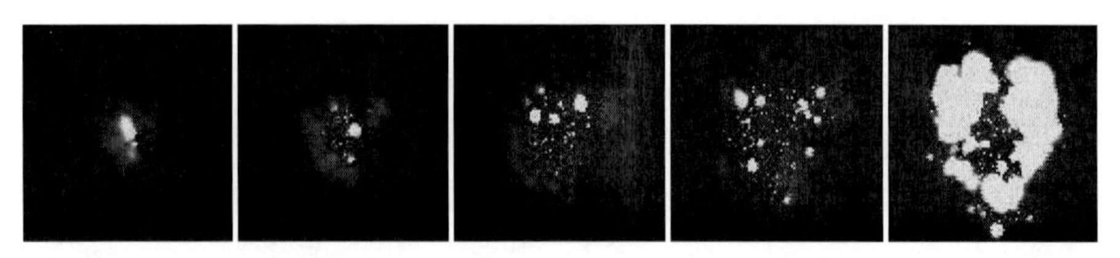

Figure 62. Five shots of a simulation of a viral infection in a lattice. Each pixel corresponds to the concentration of virus in a lattice point. Starting from a confined region, the virus spreads, mostly carried by infected cells. These burst, releasing their content and forming high-viral-density darker spots in the figure.

Colour image of this figure appears in the colour plate section at the end of the book.

five pictures shown are arbitrarily chosen among successive instants in the time sequence of the simulated infection. It is interesting to observe the development of areas with higher density of virus where infected cells have exploded and have spread their viral content. Another noteworthy aspect of this picture is the inhomogeneity in viral concentration resulting from local competition with antibodies; as in the bacterial infection example, this case shows how discrete models incorporating the 'spatial' scale are able to reproduce complex dynamics, where small differences in the local distribution of the entities play a determinant role in the outcome.

4.5 Modelling Idiotypes and the Idiotype Network

The antibody molecule expresses epitopes that are recognisable by the humoral immune system; some, in the constant regions, are markers of antibody families. Instead the *idiotypes* are in the variable regions. They are part of the paratope, and contribute to the antibody's specificity. Idiotypes have been studied thoroughly, and are considered the outer face of the paratope; a fraction of the autoantibodies, called Ab2-beta, bear similarity to the antigen epitope and are known as antigen's 'internal image'. The *idiotype network*, including Ab1, Ab2, Ab3 and so on, has been theorised by Niels Jerne in 1974 [11], as the internal generator of the entire immune repertoire. The theory elicited initial enthusiasm but postulated that the entire network was naturally sustainable, and the postulate has not been upheld.

This point is argued in 1.26; the idiotype network is not feasible unless the natural regulation of "useless" responses is lifted (e.g., by adjuvant stimulation). Even so, for modellers to create a full-blown construction, that in nature, remains limited to the first steps, is perfectly acceptable if it is acknowledged that the modellers' purpose is simply an exercise, not a reflection of nature. It is also interesting as a *'what-if'* experiment and important as a benchmark test for any model.

By enabling the normally inactive interaction between BCRs and antibodies, one can simulate the idiotype network in C-ImmSim. The experiment consists of injecting the priming antigen into the model, configured so as to bypass the requirement of a *damage signal* for the activation of helper T-cells. A chain of recognitions follows with the generation, in turn, of antigen-specific antibodies (Ab1), and also antibodies that recognise Ab1 (Ab2). Eventually, the same mechanism produces Ab3, and so on [125]. This model experiment generates a cascade of immune responses whose kinetics and specificities can be studied, as shown in Figure 63. The antigen injected at time zero has epitope 0, i.e., the binary string (bit-string length NBIT = 12) is 000000000000. Since the minimum match $m_0 = 11$, the antibodies potentially responding to the Ag are 13 (i.e., one perfect match plus twelve 1-bit mismatches). In effect, the first antibody to be produced is Ab1 = 4087 (in red, in Figure 63). Ab2 = 8 (in blue) is an anti-Ab1, and very similar to the antigen (i.e., an internal image). The clearest difference between these computer results and the hypothesised idiotype network (based on the diversity of the immune system) is that in the former all Ab2 or Ab2-like are very similar to the antigen's epitope, while in nature any Ab2 (or Ab4, etc.) response produces a small minority of internal image-carrying antibodies (Ab2-beta). Besides, all the Ab2-alpha (directed to non-Ag-binding parts of Ab1's receptor) do not contribute to the expansion of the network but increase the level of fog in the system.

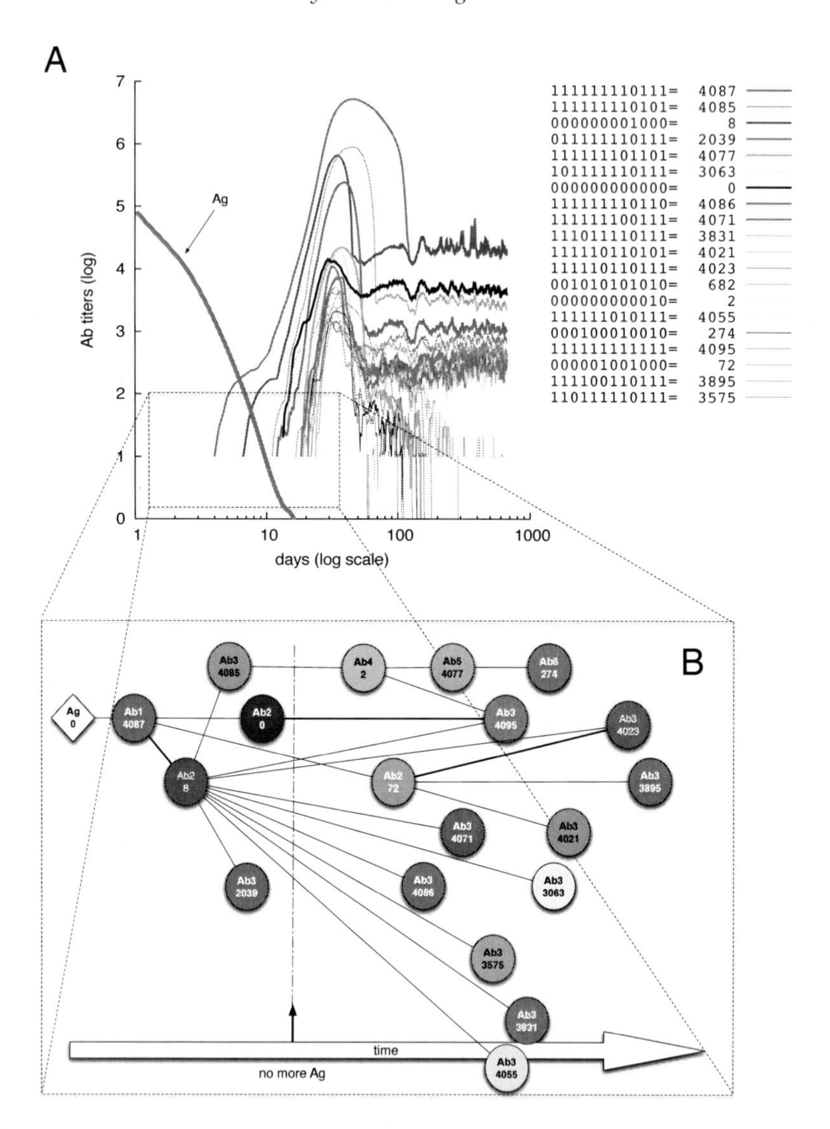

Figure 63. The idiotype network is clearly at work in this simulation result. Only the perfect match and one-bit mismatches are allowed (i.e., $m_0 = 11$). A chain of 190 idio-clones has been generated from the injected (antigen) but for simplicity, only the top twenty are shown, namely those reaching higher titers during the simulation. Some of them go to zero within about three months, whereas the remaining reach a metastable state in which they stimulate themselves, one with another. In panel B, is shown the matching relationship among the bit-strings of the idio-clones plotted in panel A. An edge between two nodes stands for a bit-string affinity above the threshold m_0. Bold edges indicate 12-bit matches and the rest, 11-bit matches. Idiotypes are shown in sequential order of appearance (left to right).

Colour image of this figure appears in the colour plate section at the end of the book.

In the present simulation, from the initial antigen injection, a total of 190 antibody clones were generated via stimulation by idiotypes; Panel A shows only the foremost twenty, where foremost means reaching higher titers. Note that late after the antigen clearance at day ~10, some of the clones go to zero in about three months, whereas others reach a metastable state. Panel B shows the bit-string affinity relationship among the clones. An edge between two nodes stands for a bit-string affinity above the threshold m_0. For example, Ab1 = 4087 stimulates Ab2 = 8 and only later Ab2 = 0 with a non-perfect match, whereas Ab2 = 0 induces the production of a later-appearing Ab3 = 4095 with a 12-bit match. Actually it is interesting to note that the major part of the network is built on a low-affinity relationship among the *idio-clones*.

Also of interest to note is that whereas most of the idio-clones decline at a given time point, a small group (about 0.7% of the potential repertoire of 2^{12} = 4096 and 14% of the total network elicited composed by 190 idio-clones) feed one another and somehow keep the immune response alive for a virtually infinite period of time.

Another interesting thing to note is that when the B-Ab interactions are enabled (and, not to forget, considerable help to the injected primer antigen is provided in the form of an adjuvant), the humoral response is much larger (Figure 64). Also, in plot B of Figure 63, Ab1 = 4087 accounts for the totality of antibodies when B-Ab bindings are not allowed (black dots) and in the other case, the relative fraction of Ab1 decreases a bit after the initial peak but still represents the 'dominant' idiotype of the network.

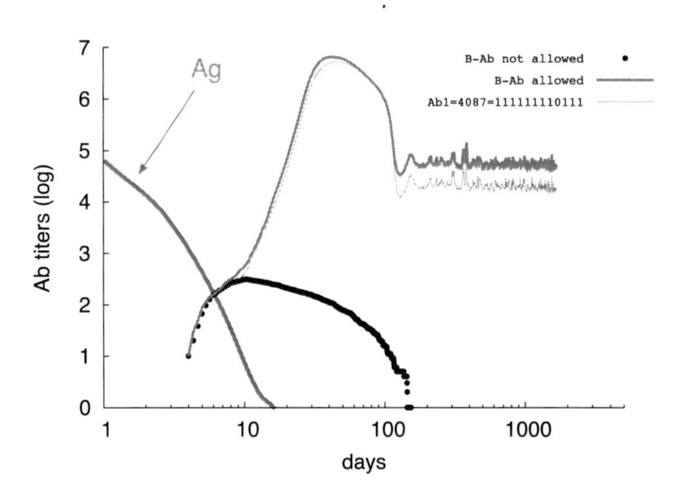

Figure 64. When B-Ab interactions are enabled, the humoral response is greater (thick orange line). Ab1 = 4087 (thin orange curve) is the dominating clone in both cases although he is just one out of 190 of the total idiotype network. The black curve shows the titer of the same antibody in case the network is not elicited. The whole humoral response consists of the Ab = 4087 only.

Colour image of this figure appears in the colour plate section at the end of the book.

5
Specific Applications

The versatility of the model becomes useful when trying to model a specific pathology such as, for example, infection by a complex virus, allergies or cancer. The modus operandi is always the same. First, identify the key players as agents, and then translate immunological knowledge about their behaviour into logical rules. Finally, after having dealt with all the technicalities of the implementation, validate the model by comparing the outcome with experimental data, keeping in mind that some simplifications and approximations are needed even when interpreting the model output.

The following sections describe the experience gained by using the C-ImmSim model to reproduce the emergence of type I hypersensitivity in atopic individuals, to describe the evolution of HIV-1 and the way it manages to survive the constant antagonism of the immune defences, and finally—in a task to which the model is particularly well-suited—to test *in silico* vaccination schedules in cancer immunotherapy.

5.1 A Multi-scale Approach to Model Hypersensitivity

Type I hypersensitivity, or allergy, consists of an overreaction to what is generally called an *allergen*. The term *allergen* indicates a molecule or set of molecules taken together that put the immune system in alarm, although by their nature they are innocuous. Allergens are treated as dangerous antigens by the immune system, which mounts an immune response to eliminate them. In healthy people, the immune system correctly ignores this 'fake threat', whereas in some individuals, a cascade of events takes place starting with the recognition of the allergen and ending with the release of histamine [91]. Histamine is what provokes all the known symptoms of allergy (itching, sneezing, wheezing, swelling), sometimes culminating in anaphylactic shock.

The term *atopic* refers to people who overreact to *some* allergen. For reasons that are not yet understood, these people have a predisposition to respond to a specific environmental antigen (e.g., pollen, mold spores) by producing antibodies of the IgE class. Since this trait tends to run in families, it probably has a genetic component. It is estimated that more than 30% of the world population is atopic. Moreover, the number of people suffering from atopic diseases is increasing in the industrialised countries, revealing a link between modern life and atopy [126].

In recent decades, scientists, clinicians and epidemiologists have elucidated the cellular and intracellular mechanisms involved in allergic reactions, including the roles of T helper subsets and interleukins. Still unknown, however, are important details such as, for instance, the full sequence of events involved in the development of the disease and the key factors determining the differences between a person who is allergic to, say, grass pollen and one who is allergic to bee venom. The only agreement seems to be that atopy is a consequence of a complex series of interactions involving not only the allergen, but also the dose, the sensitising route, sometimes an adjuvant, and most importantly, the genetic constitution of the individual.

There is no doubt that modelling allergies is an intricate task that, in principle, should embrace different levels of biological organisation, going from the gene to the cell or tissue level, through the complicated machinery of cell signalling. By using

C-ImmSim, a model has been constructed, which takes into account all these levels (Figure 65).

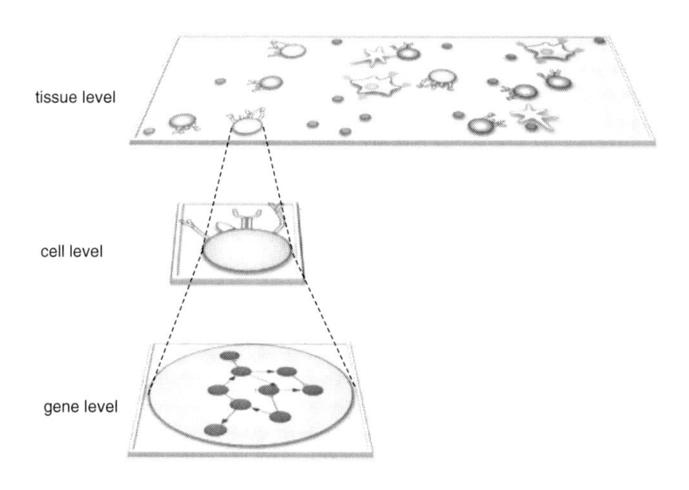

Figure 65. A multi-scale model can be constructed by simulating the gene expression dynamics by means of a Boolean gene-regulatory network coupled with the extracellular level of cytokine signaling. The cross-talk among cells represents tissue level dynamics.

Colour image of this figure appears in the colour plate section at the end of the book.

Key to understanding the pathogenesis of allergies is the differential role of helper T lymphocytes of type 1 or 2. An unbalanced differentiation of T helper cells from precursor type Th0 to the Th1 or Th2 phenotype in immune responses often leads to a pathological condition. To set out the differentiation of T helper lymphocytes, the C-ImmSim agent-based model has been integrated with a gene regulatory network model [127]. In so doing, it has been possible to determine the cells' individual differentiation fate as a function of the environmental signals (i.e., cytokines).

As usual, the model is built by translating immunological knowledge into suitable rules to be added to the automaton: (i) the allergen triggers a primary Th2 response (biased, in atopic individuals) which consists in a larger production of Th2 cytokines; (ii) one of such cytokines, IL-4, induces isotype switch in B-cells while inhibiting Th1 activation (Th1 and Th2 are antagonists); (iii) plasma B-cells developing from stimulated B-cells produce antibodies of the IgE type; (iv) IgE binds to Fc receptors on mast cells (sensitisation phase); (v) IgE molecules bound to mast cells bind newly encountered allergen (later reaction phase) and cross-link Fc-IgE complexes; (vi) cross-linking on mast cells causes degranulation and release of histamine (among other active substances).

Modelling-wise, the interesting part is the mode of linking the gene regulatory level to the cellular level: the gene regulatory network (GRN) describes intracellular-level phenomena that determine, at the upper (mesoscopic) level, the phenotypic T-cell differentiation into Th1 or Th2. The GRN used represents one of the most extensive attempts to model the regulatory network controlling the differentiation of Th lymphocytes to date [128]. Seventeen genes involved in T helper differentiation make up this network.

The first step in the model construction consisted of identifying the genes of the network coding for membrane receptors and those coding for soluble molecules to be secreted by the cell, with the idea of interpreting the former as the 'input' and the latter as the 'output' of the cell. Then the network's Boolean dynamics was analyzed using classical logical methods to identify the asymptotic regimens. In particular, three 'attractors' (in mathematical terms) with relevant biological meaning were identified, two leading to Th1 and one to Th2 phenotype. For each time step of the simulation, each undifferentiated T helper cell would individually transduce the input signals coming from the extracellular space, through the cell receptors into a micro-dynamics of the gene regulatory network, eventually falling (or not) in one of the attractors. In the case that one of the possible attractors was reached, then the rule would be 'fired' and the cell would become a Th1 or Th2, otherwise the cell would remain in the undifferentiated state.

This example shows that an integration of the two levels of description (intracellular and extracellular) can be achieved, assuming that (i) the intracellular gene regulatory network is biologically sound and allows for relevant asymptotic regimens and (ii) the stable dynamics at the lower level can be rationally translated into an action (the rule) at the upper level. It suggests a general procedure to construct a multi-scale model accounting for, altogether, gene expression data, cytologic profiles and cytokines data.

5.2 HIV Infection and AIDS

The human immunodeficiency virus or HIV (rightmost picture of Figure 56) is the cause of the nefarious Acquired Immune Deficiency Syndrome (AIDS). The features that make HIV very effective in replicating and surviving in infected human hosts are the high replication rate, the high mutation rate, and the ability of a glycoprotein expressed on its surface (gp120) to bind the CD4 receptor expressed on the surface of helper T lymphocytes [3]. The latter implies that HIV infects the component of the immune system that is responsible for the activation of both humoral and cell-mediated immune responses. Undermined on its very own grounds, the immune system is eventually defeated by opportunistic diseases, which would be wiped out in normal conditions. The fundamental CD4 T lymphocyte depletion process is likely caused by a combination of destruction of mature CD4 T-cells and impaired production of new T-cells. Biological studies (both *in vitro* and *in vivo*) have identified a number of specific mechanisms [129], including accelerated destruction of mature CD4 T-cells due to viral proliferation, chronic activation leading to apoptosis or anergy, impaired production of new T-cells by destruction of haematopoietic progenitor cells, and damage to lymphoid organs where these cells reach maturity [101]. However, to date, the relative contribution of each to the global process of T-cell depletion is unclear.

With respect to other known diseases caused by viruses, the natural history of HIV infection and the onset of AIDS shows a singular time evolution that has particularly stimulated mathematicians [130]: (i) the first contact with the virus induces a normal primary immune response, very similar to any other infection, which lasts no more than a couple of months; (ii) this acute phase is followed by a period of latency (from two to eight years) during which the HIV plays 'hide-and-seek' with the immune system; (iii) when the CD4 T-cell count drops to about 20% of the normal value (around 10^3 cells per *ml*), there is the onset of a highly immunodeficient status. At this time, the immune system becomes unable to defend its host from opportunistic diseases and the patient dies within two to three years.

So far, few mathematical models have been able to reproduce the two time scales of the AIDS development (fast primary infection or *acute* phase, and slow asymptomatic or *chronic* phase). How the C-ImmSim model has been successful in reproducing the time scales and many aspects of the immune response to the HIV is illustrated in the following figure. In particular, it matched (i) the dynamics of T helper cell depletion, both by direct destruction caused by the virus and indirect by impairing naïve CD4 T production; (ii) the effects of the formidable escape factor provided by transcription errors during HIV assembly inside infected cells; (iii) the two time scales in the disease progression; and (iv) the distribution of the 'time to AIDS' as observed in legions of field studies.

One way to show the modification in the dynamic rules of the agents' behaviour is to provide the stochastic finite state machines of the entities Th, MA and Dc (this formalism has been introduced in section 3.2.1). Helper T-cells are the ones mainly impaired by the virus, because they are infected and act as reservoirs for the virus. The virus also infects dendritic cells and macrophages. For the other cellular entities, no changes are required.

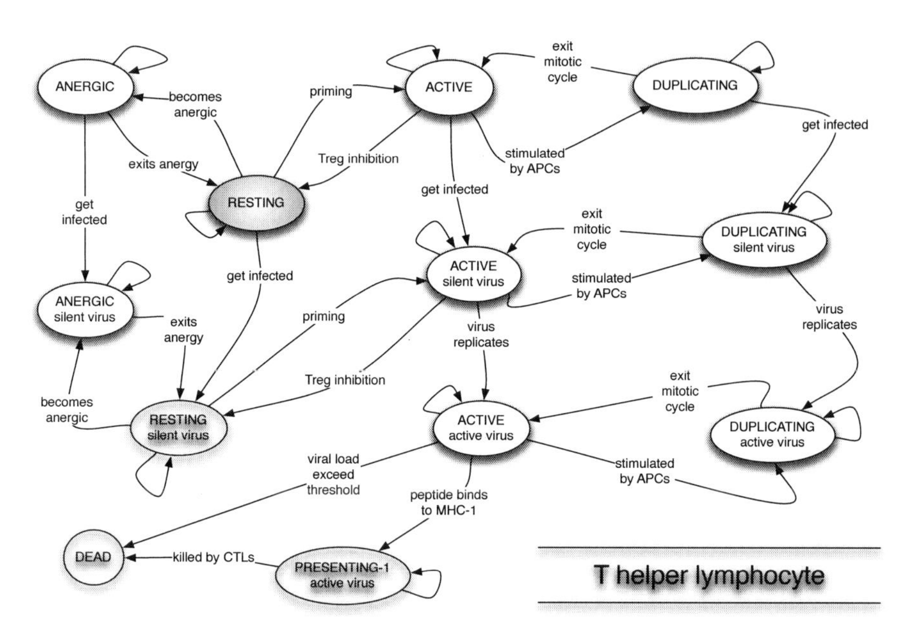

Figure 66. The HIV-1 virus infects lymphocyte T helpers by binding the gp120 molecule on its surface to the CD4 receptor on the Th cell surface. Then its viral content is inserted into the host DNA ('silent virus' represents this state). Only when the genes of the virus are transcribed ('active virus') does the number of copies of fully assembled virions increase and the cell can present the viral peptide on the class-1 MHC molecule. In this state, the Th becomes the target of cytotoxic cells recognising the peptide. The life cycle of the virus intersects those of the cell so the number of states of this SFSM is multiplied with respect to Figure 28.

It is interesting to point out that the infection of each cell type has specific consequences in the simulation. For instance, if HIV ignores the dendritic cells, the only effect in the dynamical evolution of the disease is a reduction of the cytotoxic activity in recognising new strains of HIV. By contrast, extending the infection to macrophages has a more striking impact since it weakens the innate response during the first phase of the infection. Moreover, it partially reduces the number of active antigen-presenting cells, lowering the efficiency in stimulating helper cells and thus impairing both cytotoxic and humoral responses.

Cell infection is modelled as a stochastic event, that is, there is a fixed probability that an HIV particle infects a target cell. Once

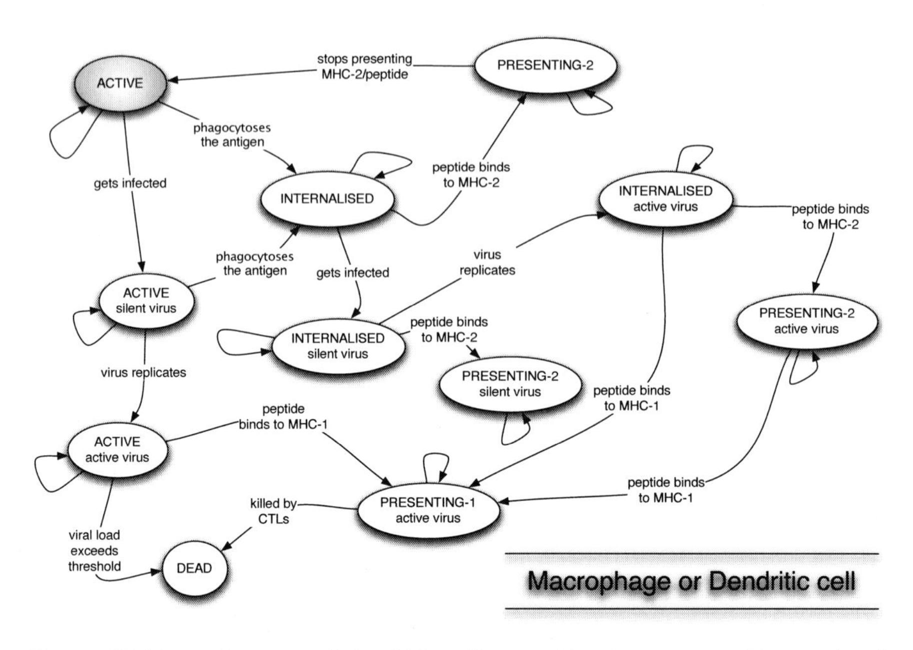

Figure 67. Macrophages and dendritic cells are also virus targets. The stochastic finite state machine representing their dynamical rules is more complex than before (Figure 27 and Figure 30).

inside the target cell, the virus transcribes its RNA genome in the host DNA. Then the virus remains silent until a subsequent stochastic event triggers viral genome transcription. It follows viral assembly and proliferation. After that, fully assembled viruses start budding from the cell membrane and reach the outer extracellular space.

Two binary strings code for the virus (each NBIT bits long); one corresponds to the epitope (i.e., the BCR's binding site) and the other to the peptide (i.e., the MHC class-1 and class-2 binding sites). In the simulations shown below, the bit-string length NBIT is equal to 12 for a potential repertoire of 4,096 distinct receptors, epitopes and molecules. Actually, since HIV is represented by one epitope and a peptide, 24 bits identify each virus, so the potential number of different virus strains is equal to 16,777,216.

Each viral strain is characterised by a triplet of numbers, p_b, p_w and p_r, indicating respectively the mutation, activation and replication rate. Since mutation was allowed to modify each one of these parameters, the evolution of the population of virus followed a Darwinian selection process in all respects.

The *per-bit* virus mutation probability, p_b, is taken equal to 0.01, which means that the probability of having at least one mutation in the string defining the virus is $1 - (1 - p_b)^{24} \sim 0.22$ (in accordance with [23]). To take into account the virus transcription errors, the two bit-strings undergo bit-flips according to the following simple scheme: first the model selects, with equal probability, either the peptide or the epitope as a candidate for the mutation, then flips a randomly chosen bit with constant probability, p_b. If the mutation affects the epitopes (i.e., it is an *antigenic shift*), the new viral strains induce a different response from the humoral branch of the immune system. Otherwise, if the peptide is mutated, it is the cellular response that is affected.

The initial number of cells scattered on 400 grid points has been set equal to about 500 per *ml* for B lymphocytes and target cells (macrophages and dendritic cells), whereas the number of cytotoxic T cells is twice as many [91]. Helper T-cells are set to about 1,500 per *ml*. As usual, each time step of the simulation corresponds to eight hours of real life.

Viral DNA's transcription in infected cells is activated with a certain probability, p_w, called *activation* or *transcription rate*. During the asymptomatic phase, the immune response causes a selection such that only those variants of the wild type having a 'suitable' value of the activation rate have good chances to survive and proliferate. Escape-viral mutants with a low transcription rate should be selected during the silent asymptomatic stage of HIV disease. On the other hand, the selection of escape-viral mutants with a very low transcription rate would generate a permanent silent/asymptomatic stage, as present in a minority of HIV-patients, called long-term non-progressors. As a consequence, there should be a range of intermediate values for the activation rate such that, on average, the corresponding time delay with respect to the cell infection is optimal from the HIV's point of view. To account for such an important feature, the activation probability, p_w, in the model is unique for each viral strain and the corresponding value is stochastic in turn. The optimal value of p_w is automatically determined by selection, just as it is likely to be in real life, where the delay depends on both the virus' genome itself and the metabolic activity of the infected cell. The emergence of the latent phase appears as the result of the selection of viral strains, which code for a low transcription rate, p_w, thus stretching the time between the infection of the cells and their activation.

When simulated in the computer, this dynamics shows interesting features. Driven by mutation and selection, the viral strain's population evolves under the selective pressure of the immune system. On its part, the latter struggles with the ever-changing virus which manages to escape immune control, thanks to the high rate of mutation. The result of this relationship can be plainly seen in Figure 68 where the virus population is shown as a function of the number of mutations from the wild type virus (i.e., the viral strain that originally infected the virtual host at the beginning of the simulation). The distance is expressed in terms of bit mutations both on the epitope and the peptide for a total of 2 × NBIT = 24 possible mutations. The viral count of the wild type and its offspring has many peaks, but two are really prominent. One corresponds to the wild type itself (bit-mutation = 0), rising at about year one. Others peaks are visible at about the same time due to an early mutant that evolves within the first year. The fact that both the wild type and the mutants rise after about 1.5 years is probably due to the drop of the immune system efficiency, caused by the virus itself. Also worth noting is that the early mutants give rise to a whole 'family' of relative strains that differ from the wild type by 10 to 15 bits, meaning a 'genotype distance' between 40 and 60%. This considerable

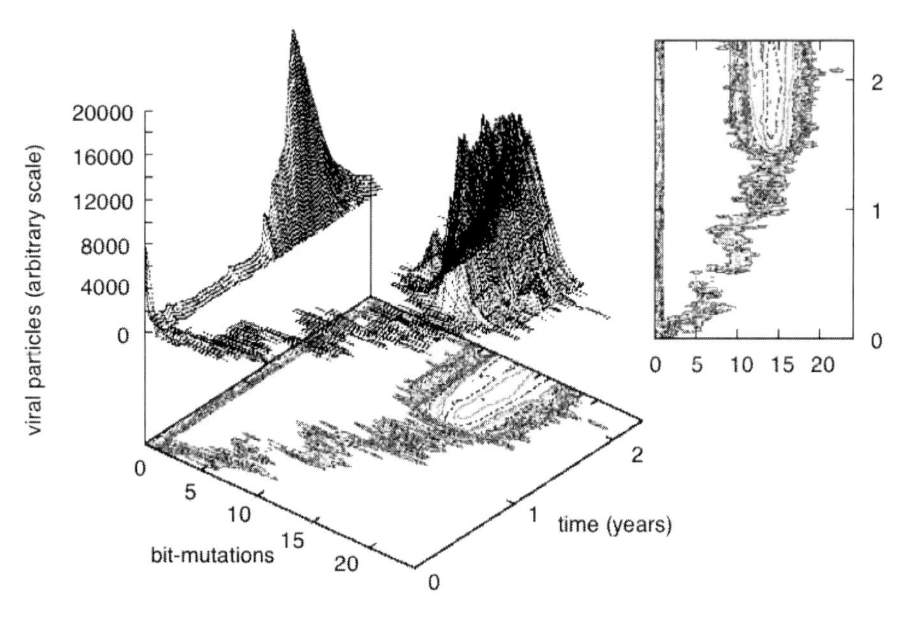

Figure 68. HIV population as a function of the number of mutations from the wild type virus. The distance is expressed in terms of bit mutations, both on the epitope and the peptide, for a total of 2 × NBIT = 24 possible mutations.

Colour image of this figure appears in the colour plate section at the end of the book.

evolution is perhaps unrealistic but gives the idea of mutation as the finest weapon of the HIV.

It is also interesting to note how the outcome of this evolutionary dynamics is totally driven by chance. In Figure 69 and Figure 70, different simulation outcomes from the same initial conditions are shown. Some of them progress to the AIDS phase in a few years (short-term progressors in Figure 69) while other have a slower evolution (long-term progressors in Figure 70). A quick progression does not necessarily mean a 'longer' evolution (i.e., the accumulation of more bit-mutations). Instead, a winning strategy for the virus appears to be avoiding engaging the immune system in 'bloody' battles until it becomes weakened. It is also interesting to note how the wild type virus sometimes manages to survive, while at other times, it succumbs to the immune system, usually within a few months or years.

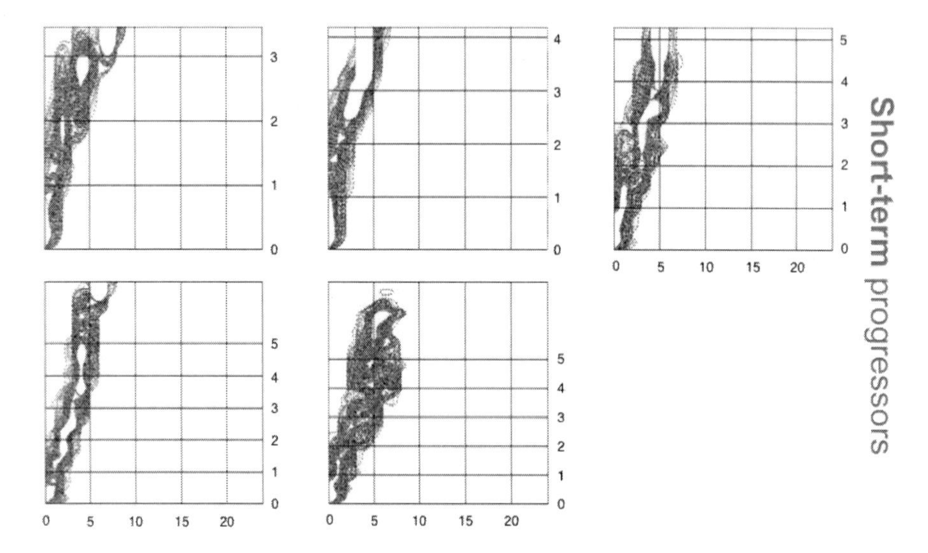

Figure 69. The evolution of the viral strains show different patterns in the short-term progressors. As in Figure 68, x-axis stands for bit-mutations and y-axis for time.

Colour image of this figure appears in the colour plate section at the end of the book.

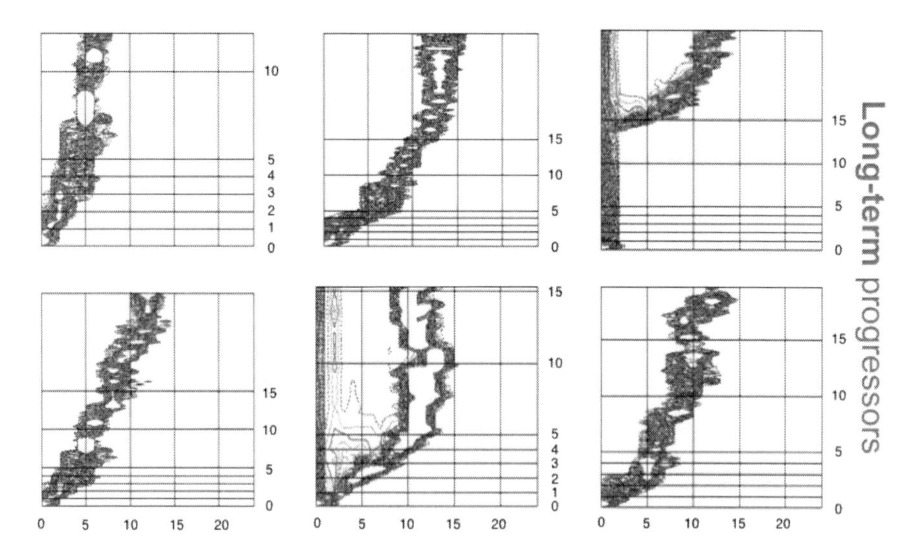

Figure 70. Long-term progressors show that the virus has a slower but more effective evolution path.

Colour image of this figure appears in the colour plate section at the end of the book.

5.2.1 Two Time Scales in the Onset of AIDS

The acute phase of the infection caused by the wild type strain is followed by the asymptomatic phase in which new strains of the virus remain silent inside infected target cells. The stochastic activation of the viral genome transcription process triggers localised proliferation, which usually does not reach high levels. However, these limited infections keep the immune system busy in the production of anti-HIV antibodies and impair the efficiency of the Th production mechanism.

Figure 71 shows the course of the total number of virions up to fifteen years from the initial infection (note that there are two different time scales). During the acute phase, the wild type virus load reaches a peak after (about) three weeks. At the same time, a primary response to the wild type virus has been developed. The response wipes out both soluble virus and virions that have started to duplicate inside infected target cells in about seven weeks. Only viruses that have silently infected target cells

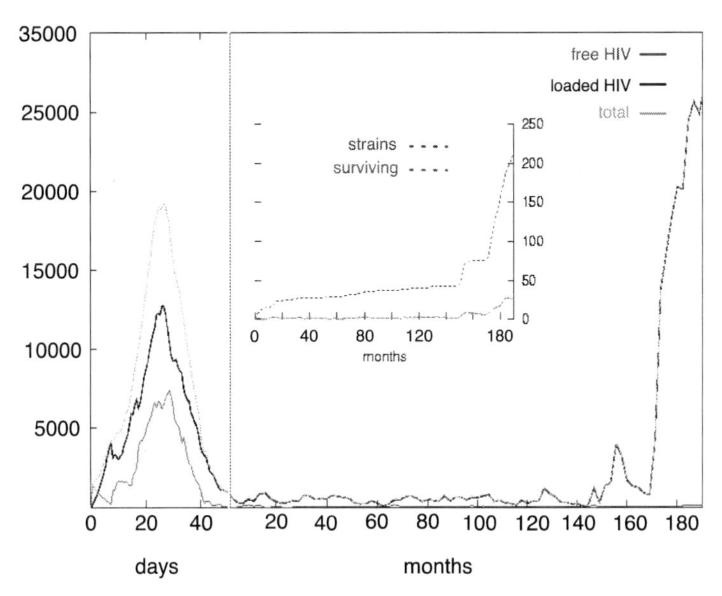

Figure 71. Virions inside infected cells (loaded) and free virions. In the inset plot are the number of generated strains and those surviving the immune selection.

Colour image of this figure appears in the colour plate section at the end of the book.

remain undetected. In principle, infected cells may expose the virus' peptides, combined with MHC class-1 molecules, on their surface. However, MHC-1 molecules bind peptides following a string-matching procedure (see Figure 42). A non-perfect match entails a much lower probability to bind, so there is a chance that virus' peptides in infected cells are not shown to cytotoxic cells. In those cells, HIV can grow undisturbed. Note that only a very small fraction (about 1%) of the virions growing inside an infected cell is fully assembled, so the number of new infections is limited. This also contributes to the permanence of the system in the meta-stable state for a long time.

The number of new strains is plotted in the inset of the same figure. Although a large number of strains are generated during the AIDS phase, few are selected for survival.

5.2.2 Helper T-cells are the Key

It is widely accepted that, in people infected by HIV, there is a deterioration of the secondary lymphatic organs [101, 129]. From this viewpoint, the onset of AIDS might seem a consequence of the decreased efficiency of the immune system productive compartments. This would mean that the immune system defeat is not due to the death of the Th cells (which is caused, in turn, by the virus infection), but mainly to the failure of the system in maintaining homeostasis.

Indeed, it has been recently shown, by means of a simple mathematical formalism, that no single direct or indirect mechanism of CD4 T-cell destruction can explain the long-term decrease in CD4 T concentration leading to AIDS. The long-term decrease in CD4 T concentration can be explained only by some positive feedback of the destructive mechanism on itself.

It is remarkable that simulations confirm such theoretical prediction: if no direct impairment of T-cell production is taken into account, then the Th cell count never reaches the critical threshold of 20% of the initial value. That is, either HIV is depleted (and the Th count is restored to the normal initial level) or it grows very fast. In such a case, the Th cell count reaches a lower-than-normal steady state but never approaches 20% (not shown).

To simulate the long-lasting, steady decline of the Th cell count it has been necessary to introduce a new mechanism which reduces the efficiency in the production/maturation of T-cells. When the number of infected cells outgrows a predefined threshold, the rate at which mature Th cells enter the lymphatic system must be trimmed (as if each revamp of the infection caused progressive damage to the lymphatic organs). The results are shown in Figure 72. In the immunodeficiency status, the capability of the immune system to re-generate diversity drops dramatically. This is evident when looking at Figure 73, which shows the distribution of T lymphocyte receptors. At the start of the simulation, there is a uniform distribution of different receptors among the Th population. In a short time, cells whose receptors have the highest affinity for the virus are stimulated and the memory is built. Receptors that are not selected by any virus strain slowly disappear since the thymus, which undergoes an efficiency reduction, does not replace them

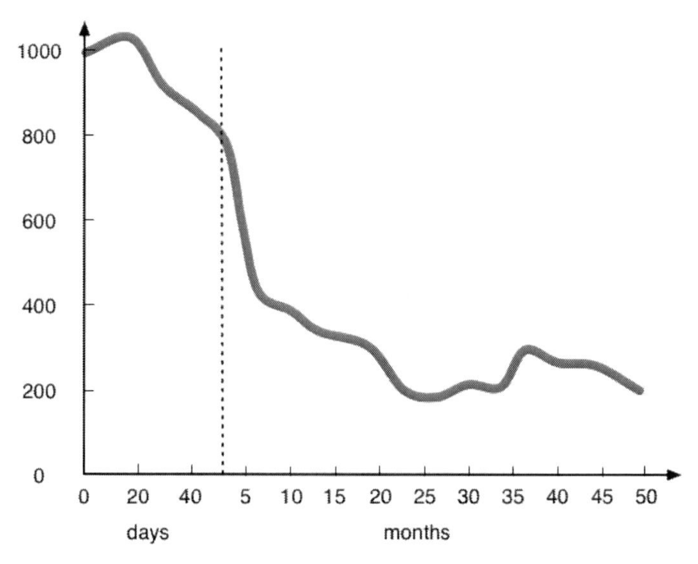

Figure 72. T helper counts. Note the two different time scales on the x-axis.

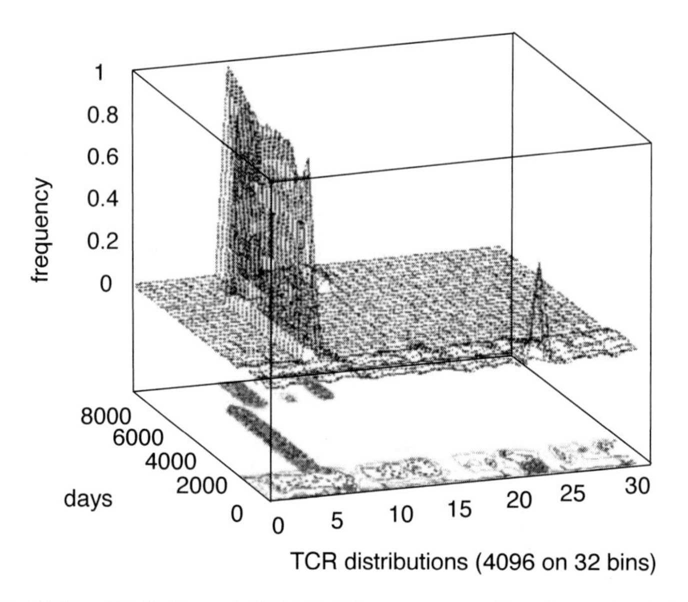

Figure 73. TCR's distribution. 4,096 bit-strings representing the potential repertoire of helper T-cell receptors have been clumped into 32 bins for clarity. Note the initial unbiased distribution quickly turning into a peak (bin~24) during the acute phase of the infection. A few memory clones (bin~7) eventually dominate the scene.

Colour image of this figure appears in the colour plate section at the end of the book.

(at a normal rate). Eventually, the immune system is weakened in its ability to generate fresh diversity; hence it becomes more vulnerable to new HIV strains and/or to other antigens.

5.3 Immunodominance in Cancer Immunotherapies

Anti-cancer vaccination is based on the existence of markers expressed by tumour cells (the so-called tumour–associated antigens or TAAs). In generic terms, cancer vaccination aims at eliciting an immune response toward one or more TAAs after injecting them together with highly immunogenic epitopes (named the *carrier*), generally derived from another organism. Cancer eradication using TAA vaccination has been demonstrated in numerous animal models. Often, however, vaccination protocols designed to elicit anti-cancer immune responses in humans have failed [131]. The high tumour burden and the compromised immune system of patients have been blamed for the clinical failures [132, 133]. Other decisions are likely to influence the outcome; for example, what antigens are used, what vaccination schedule has been adopted and what adjuvant is added to the inoculum (TAA are, by definition, self-peptides and therefore weak immunogens, unable to elicit an effective immune response).

Most clinical trials contemplate multiple injections of the TAA inoculated along with one highly immunogenic carrier. However, there is evidence that injections of two different vectors encoding the same recombinant antigen (e.g., priming with plasmid DNA and boosting with recombinant modified *vaccinia Ankara*) may generate high levels of specific immunity [134]. Of the many thousands of peptides encoded by a complex antigen potentially presented to CD8 T-cells, only a small fraction, called *immunodominant*, induces a non-negligible response in association with any given MHC class-1 allele [135].

In a situation where systematic comparison of the efficacy of different vaccination protocols is hampered by technical limitations, the use of a computational model such as C-ImmSim can help in reinforcing the evidence that the use of multiple vectors has advantages over single carrier injections [136]. The following paragraph describes the comparison of vaccine protocols that use multiple vectors/carriers and verify the effect of anti-tumour T-cells and antibody titres on the tumour.

The micro dynamics of the modified C-ImmSim model can be summarised as follows: soon after the injection of the low immunogenic TAA, together with the high immunogenic carrier epitopes, the APCs present the latter on the MHC-1 molecules for recognition by Tc cells. At the same time, the peptides are presented by the APCs bound to the MHC-2 molecule, which is then in charge of the presentation to CD4 T-cell receptors. Helper T-cells binding the complex are stimulated and release interleukin-2. IL-2 sustains the proliferation of active CD8 T-cells (i.e., by injecting the TAA, one forces the stimulation of TAA-specific CD8 T-cells). Such cells recognise the MHC-1/peptide molecules on the cell surface. This may happen on APCs but also on cancer cells. Upon recognition of the MHC-1/peptide molecule, CD8 T-cells kill the malignant cells and are, in turn, re-stimulated.

As usual, a finite state automaton is used to describe each new agent that is introduced in the model (Figure 74). In this case, the newcomer is the agent representing a malignant cell. It follows simple rules: it duplicates until cytotoxic cells recognise some peptide on MHC-1 and kill it.

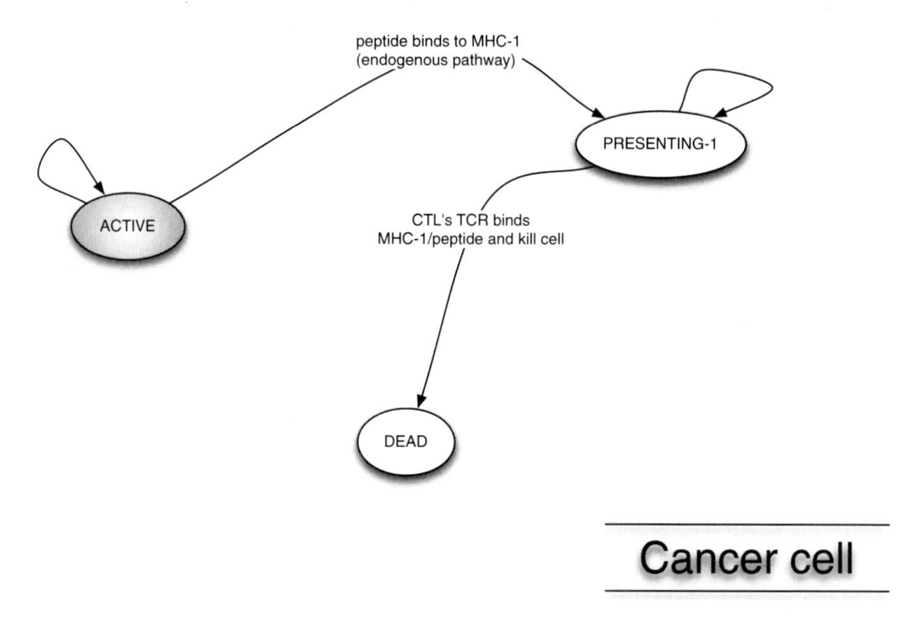

Figure 74. Cancer cells represent malignant cells that keep replicating. In the initial stage, they are considered ACTIVE (and replicating). Upon successful binding of the class-1 MHC molecule with endogenous peptides, they go into the PRESENTING-1 state. In this state, they are vulnerable to cytotoxic T-cells that can bind the MHC-1/peptide complex to their TCRs. A CTL that recognises the MHC-1/peptide complex will kill the cancer cell (DEAD state).

5.3.1 In silico Experiments

A number of ad hoc *in silico* experiments have been performed to probe whether the use of multiple vectors can boost the immunotherapy efficacy. The control case gets no treatment. In this situation, the cancer grows freely and there are few chances for the immune system to mount a cellular or humoral specific response because of the very scarce immunogenicity of TAA. Instead, when the TAA alone is administered, in accordance with a large body of experimental evidence [131], a very weak response is observed simply because of the scarce immunogenicity of the TAA. This is no surprise in C-ImmSim, since the bit-string representing the TAA is chosen as belonging to—or being very similar to—the set of bit-string representing the self. In the third scenario, the anti-cancer vaccination is realised by a combination of TAA and a carrier (C_1) identified by a bit-string purposefully chosen to be highly immunogenic. Already, in this case, an improvement in the cytotoxic as well as in the humoral response has been observed, although it was mainly directed toward C_1 rather than to the TAA (this is what is called immunodominance [137]). As the last scenario, the most interesting one, each immunisation injection uses the same TAA but combined with a different carrier molecule C_1, C_2 … C_{10}. The first interesting finding is that while the immune system mounts a different response against the carriers, it also creates the microenvironment for the recognition of the TAA. In fact a large number of active Th cells secreting IL-2 are found. Also significant is that there are fewer carrier-specific (Ab anti-C_k) antibodies, and therefore not so much competition for the injected TAA-C_k vaccine compound is observed. The net result is a larger immune response against the TAA, both cytotoxic and humoral.

To measure the effects of rotating the carrier during the vaccination schedule, two different sets of simulations have been performed. The first, labeled 'No Rotation' (NR), consists of injecting TAA+C_1 (that is, using the same carrier), whereas the second case, labeled Yes Rotation (YR), includes vaccinations with TAA+C_k with varying carrier C_k, k = 1 … 10. Moreover, the inter-vaccination time interval was varied to find the optimal schedule.

Figure 75 shows the percentage of success (i.e., virtual survivors) for the two situations (NR vs. YR) and for all injection schedules analyzed.

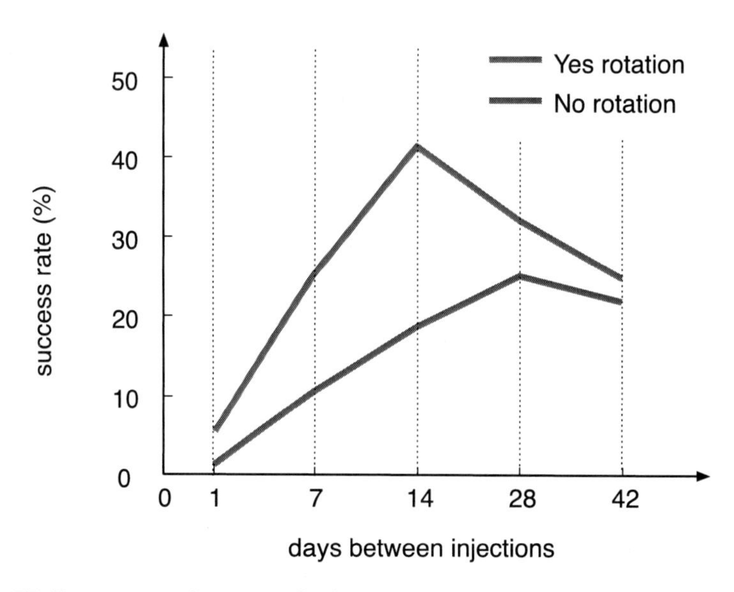

Figure 75. Percentage of success for NR and YR simulation settings for all different injection protocols tested. A successful simulation is one in which the number of tumour cells at the end of the run was lower than at the beginning. Chi-square tests indicate that for the 1, 7 and 14 days sets of simulations, the difference between NR and YR is significant (respectively 5%, 1% and 1%) whereas for the 28 and 42 days sets of simulation, it is not significant (from [136]).

Colour image of this figure appears in the colour plate section at the end of the book.

Overall, it appears that carrier rotation is advantageous. To simulate vaccination with a rotating carrier, the 14 days-interval schedule is optimal, meaning that it triggers a more robust immune response and achieves higher tumour eradication rates. The hypothesis is that the 14 days-interval is the best tradeoff between two opposite requirements: (i) higher frequency of injection that eases an early generation of the immune response, when the tumour mass is smaller and (ii) enough time between injections for the development of anti-TAA memory.

Notably, the 14 days-interval induces the highest average number of TAA-specific cytotoxic T-cells amongst all simulations (Figure 76) and, at the same time, leads to the survival of the host (Figure 75). Although delaying injections can induce a high number of anti-TAA CD8 T-cells in the long run, it is obviously less effective in eradicating the tumour.

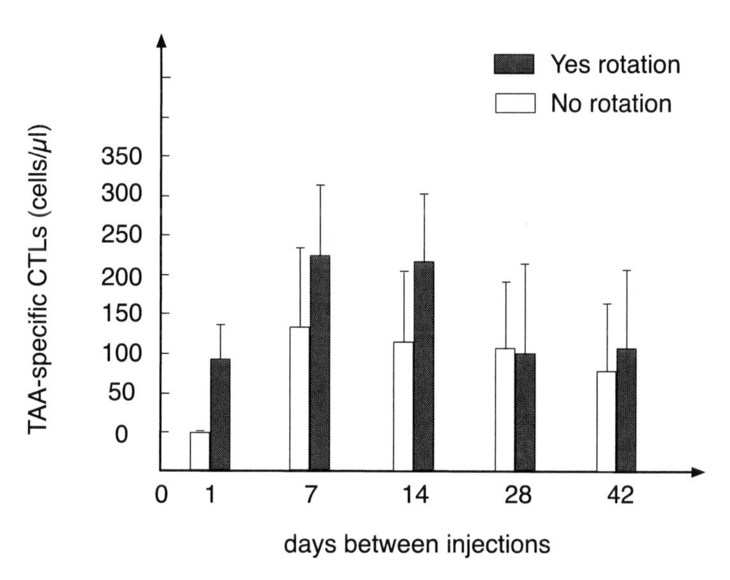

Figure 76. Number of TAA-specific CD8 T-cells generated in different vaccination protocols. CD8 T-cell numbers are computed 3 months after the end of therapy. Only simulations in which tumour eradication was observed entered the calculation.

The effect of the humoral immune response is not less important. Titres of antibodies are slightly higher (the difference is in the order of 5–20%) in the NR groups for each schedule of vaccination and this response is mainly directed toward the carrier rather than toward the TAA.

The conclusion is that in the single carrier vaccination (NR group), the generation of significantly higher titres of antibodies against the carrier is likely to be responsible, at least in part, for the reduced anti-tumour effect. Competition is what plays the pivotal role here; it is generated amongst circulating antibodies and APCs for antigen presentation of TAA molecules. Moreover, competition develops almost invariably amongst different T-cell clones. Notably, this outcome is in agreement with wet-lab experimental evidence showing that improvement of anti-TAA immunity is obtained through removal of competing carrier epitopes [138]. Also, competition among CTLs has been demonstrated to limit the immune response [139], at least in specific experimental settings.

5.4 Embedding Immunoinformatics Predictions

The use of agents to represent lymphocytes allows a step further in the representation of the immunological specificity. Indeed, a recent modification of the C-ImmSim model [140] implements a novel approach in which the molecules represented (i.e., TCRs, BCRs, antigen's epitopes and peptides) are not strings of zeros and ones but rather strings of letters representing the twenty amino acids.

This modification is a great step toward realism but carries a number of complications and requires a number of weighty assumptions. What follows is a report of this exploratory exercise, performed with the hope that, in the future, these assumptions will be removed by replacing the methods employed at present with new, more reliable techniques.

The use of amino acids in place of binary string does not bear any difficulty other than the mere technical burden of representing a variable in a set that is much larger than the one used so far. True, this step up is enormous (from 2^{16} to 20^{20} or more), but this is not the only source of complexity or of approximation. Other points are: how to determine the affinity between amino acid strings, how to calculate the likelihood that a peptide is shown in the context of class-1 and 2 MHC, and how to determine what the peptides are. The answers are very complex and they are at the core of a new scientific area of research, called *immunoinformatics* [141].

Immunoinformatics is a new, experimental and theoretical discipline emerging from the growing knowledge gathered for decades in experimental immunology and immunogenomics [142, 143]. In immunology, what is of utmost importance is to 'predict' which part of the antigen will constitute an immunogenic epitope and, broadly speaking, there are two ways of doing it. The first is to simulate the chemical-physical interactions between peptides and MHC molecules (e.g., Nanoscale Molecular Dynamics or Adaptive Biasing Force methods [144]), which takes hours to simulate a single interaction. The second way is to resort to bioinformatics approaches that use *machine learning* and *statistical methods* to extract and generalise information from available *experimental data* of, for instance, MHC-peptide sequences (for a review see e.g., [145]). This is what is intended for immunoinformatics [146].

Peptide discovery is one of the areas of most active research. MHC class-1 binding prediction methods based on machine learning have increased their accuracy over the years, finally leading to 'reliable' predictions [147, 148]. The same level of accuracy has not yet been reached by class-2 peptide prediction methods [149]. This is thought to be due to the open structure of the binding pockets in class-2 molecules, which allows peptides of different lengths to bind to the groove. In contrast, class-1 molecules restrict the size of the peptides they bind to 8–12 amino acids [135, 150] with an average length of 9 amino acids.

In order to simulate an immune response to a pathogen, from the point of contact with the antigen to the development of both humoral and cytotoxic responses, one needs a reliable method to predict class-1 and 2 peptides of the antigen (which exist), and in addition (i) a method to predict B epitopes in the antigen amino acid string, and (ii) a method to assess the binding affinity between T-cell receptors and MHC/peptide complexes.

The majority of tools available for the prediction of 'linear' B-cell epitopes are based on approximated *propensity scales*. These methods assign a propensity value to each amino acid in the queried protein sequence based on knowledge of the amino acid's physical and chemical properties. Propensity scales have been developed based on antigenicity, hydrophilicity, inverted hydrophobicity, accessibility, and secondary structure. The propensity scales of Parker (based on hydrophilicity) [151] and Levitt (based on the secondary structure) [152] show better performance in comparison to others.

As far as the assessment of the probability of binding among cell receptors and generic molecules is concerned (e.g., epitopes, MHC/peptide complexes), one method, employed in [140], uses a general protein-protein binding potential on the basis of the work of Miyazawa and Jernigan on protein energy potentials [153-155]. This method should assess the chances of direct interactions among proteins. A protein–protein potential is derived from the analysis of 3D structures in which the relative positions of amino acids are determined; the contact potential matrix reflects the decrease of entropy produced by the binding between two residues.

5.4.1 A Critique and Some Encouragement

To produce reliable simulation results, all the steps listed in Table 9, necessary to mimic the development of an immune response, have to guarantee an *unusually high* level of reliability in their prediction because the level of trust of the final result depends on the precision of each single task. The overall reliability, R, can roughly be estimated (assuming a linear dependence between the accuracy of each sub process) by this formula: $R = A_E \times A_M \times A_T$ where A_E is the accuracy of the B epitope prediction, A_M is the accuracy of MHC (either class-1 or class-2) peptide prediction and A_T is the accuracy in assessing the affinity (hence the binding) between a TCR and a MHC/peptide complex. Just to give an idea of what could come out, an accuracy of 80% (that is not generally considered bad) in each of these steps would determine an overall accuracy of the prediction of the immune response of about 51%. That is slightly better than deciding by flipping a coin.

Table 9. Immunoinformatics methods used in C-ImmSim to compute epitopes, peptides and affinity among molecules in a comprehensive model of the immune response.

What	**Notes**	**Method employed in** [140]	**References**
B-cell epitopes prediction	Most prediction tools for B-cell epitopes are based on linear predictions and use propensity scales. These methods assign a propensity value to each AA in the queried protein sequence based on knowledge of the AA's physical and chemical properties.	The propensity scales of Parker (based on hydrophilicity) and Levitt (based on the secondary structure)	[150] [151] [152]
Class-1 and -2 MHC prediction of peptides	Class-1 peptides predictions show a better performance compared to class-2 predictions.	Position Specific Scoring Matrix (PSSM)-based method	[145] [156] [157] [158] [159] [160] [161]
Binding among BCRs and Ag-epitopes, TCRs and MHC-peptides complexes.	There are no prediction tools available for this. A generic contact potential among amino acid sequences has been proposed. The work performed by Miyazawa and Jernigan on protein energy potentials provides a rough method for assessing the chances of direct interactions among proteins in the simulation.	The contact potential of Miyazawa and Jernigan	[153] [154] [155]

Not really useful.

So why bother? The answer can be formulated by adopting a different point of view or by seeing it in perspective. The novelty of this approach lies in the use of a combination of the agent-based model and of molecular binding prediction methods. As shown in [140], the combination of these algorithms allows performing *in silico* experiments with specific real-world proteins (or at least their images), which does not require extrapolating the results from an abstract mathematical space. Now the point is that techniques employed to implement these two independent sets of algorithms have shown great improvements in the past. Thanks to the ever-growing availability of fast computers, computational models are not limited *per se*. As far as the strength of the immunoinformatics prediction tools is concerned, it must be said that a great deal of enhancement has been attained in the past decade and there are reasons to expect further improvements in the years to come [162-164]. Will this allow systems to answer questions about the *specificity* of an immune response to pathogens?

To conclude on a positive note, this so-far-unique system provides a framework for testing various prediction methods, because the two levels of description, the molecular and the cellular, are quite independent. This disentanglement facilitates testing of alternative models; thus, for example, a novel method for predicting B-cell epitopes can be easily 'plugged' into the simulator, and its value assessed by observing the resulting immune dynamics.

This could be useful.

6

One Last Word

The conclusion of this book has finally been reached. We have recounted an interdisciplinary adventure spanning about two decades, an adventure about a pioneering idea which has inspired friends, immunologists and aliens. But the story continues.

Immunologists show new faces of immune phenomena. Imaging is producing incredibly detailed information about cell behaviour *in vitro* and *in vivo*. Genomic and proteomic data are literally overwhelming. Even so, the task of disentangling the complexity of the immune system remains challenging. Observed from different perspectives, the immune system appears as an impervious castle. It is characterised by a vast combinatorial diversity, by the local nature of certain triggering events linked to the systemic reach of their consequences, by the time/space multi-scale quality of its action, by the redundant and multivariate amount of its signals, etc. All this literally baffles researchers, hides the evidence, complicates the theories, and suggests exceptions. In diseases, all the complications act together, making up a giant wall, impossible to overcome, difficult to penetrate.

There is a multidisciplinary army laying siege to the castle. Aliens formulate *weapon* models that are more and more efficient and predictive; they devise *siege-machine* computers that are increasingly powerful. Immunologists rejoice at the new paraphernalia of weapons and work extra hours to provide those *in vitro* and *in vivo* experimental validations needed by aliens' models in order to keep the new Science anchored to sweet Earth and sweet Nature.

The story continues.

7

Bibliography

1. Celada F, Seiden PE. A computer model of cellular interactions in the immune system. Immunol Today 1992; 13(2): 56–62.
2. Szilard L. Report on 'Grand Central Terminal'. New York: Lyle Stuart, Inc.; 1949.
3. Manca F, Seravalli E, Valle MT, Fenoglio D, Kunkl A, Li PG, et al. Non-covalent complexes of HIV gp120 with CD4 and/or mAbs enhance activation of gp120-specific T clones and provide intermolecular help for anti-CD4 antibody production. Int Immunol 1993; 5(9): 1109–17.
4. Lefèvre O, Seiden PE, Celada F. Insights into rheumatoid factor production using a cellular automaton model of the immune system. Internat J Appl Sci Comput 1993; 3: 32–47.
5. Lanzavecchia A. Antigen-specific interaction between T and B cells. Nature 1985; 314(6011): 537–9.
6. Howard JC. Immunological help at last. Nature 1985; 314(6011): 494–5.
7. Celada F, Seiden PE. Affinity maturation and hypermutation in a simulation of the humoral immune response. Eur J Immunol 1996; 26(6): 1350–8.
8. Berek C, Milstein C. Mutation drift and repertoire shift in the maturation of the immune response. Immunol Rev 1987; 96: 23–41.
9. Burnet FM. The clonal selection theory of acquired immunity. London: Cambridge University Press; 1959.
10. Jerne NK. The immune system. Sci Am 1973; 229(1): 52–60.
11. Jerne NK. Towards a network theory of the immune system. Ann Immunol (Paris) 1974; 125C(1-2): 373–89.
12. Matzinger P. Tolerance, danger, and the extended family. Annu Rev Immunol 1994; 12: 991–1045.
13. Hozumi N, Tonegawa S. Evidence for somatic rearrangement of immunoglobulin genes coding for variable and constant regions.

Proc Natl Acad Sci USA 1976; 73(10): 3628–32.
14. Porter RR. Lecture for the Nobel Prize for physiology or medicine 1972: Structural studies of immunoglobulins. 1972. Scand J Immunol 1991; 34(4): 381–9.
15. Edelman GM. Antibody structure and molecular immunology. Scand J Immunol 1991; 34(1): 1–22.
16. Oudin J, Michel M. Une nouvelle forme d'allotypie des globulines γ du sérum de lapin apparemmentliée à la fonction et a la spécificité anticorps. [A new allotype form of rabbit serum gamma-globulins, apparently associated with antibody function and specificity]. C R Hebd Seances Acad Sci 1963; 257: 805–8.
17. Kunkel HG, Mannik M, Williams RC. Individual Antigenic Specificity of Isolated Antibodies. Science 1963; 140(3572): 1218–9.
18. Lindsley DL, Odell TT, Jr., Tauchse FG. Implantation of functional erythropoietic elements following total-body irradiation. Proc Soc Exp Biol Med 1955; 90(2): 512–5.
19. Varela FJ. Structural coupling and the origin of meaning in a simple cellular automaton. In: Sercarz EE, Celada F, Mitchison NA, Tada T, editors. The Semiotics of Cellular Communication in the Immune System. Berlin and Heidelberg: Springer-Verlag; 1988, pages 151–61.
20. Von Neumann J. Theory of self-reproducing automata. Urbana, IL: University of Illinois Press; 1966.
21. Wolfram S. Cellular Automata and Complexity: Collected Papers. Reading, MA: Addison-Wesley; 1994.
22. Gardner M. Mathematical Games—The fantastic combinations of John Conway's new solitaire game "life". Scientific American 1970; 223: 120–3.
23. Celada F, Seiden PE. Modeling immune cognition. IEEE International Conference on Systems, Man and Cybernetics 1998; 4: 3787–92.
24. Celada F. Does the human mind use a logic of signs developed by lymphocytes 10^8 years ago? In: Sercarz EE, Celada F, Mitchison NA, Tada T, editors. The Semiotics of Cellular Communication in the Immune System. Berlin and Heidelberg: Springer-Verlag; 1988, pages 71–9.
25. Lanzavecchia A. Receptor-mediated antigen uptake and its effect on antigen presentation to class II-restricted T lymphocytes. Annu Rev Immunol 1990; 8: 773–93.
26. Steinman RM, Witmer-Pack M, Inaba K. Dendritic cells: antigen presentation, accessory function and clinical relevance. Adv Exp Med Biol 1993; 329: 1–9.
27. Zinkernagel RM, Doherty PC. Restriction of *in vitro* T cell-mediated cytotoxicity in lymphocytic choriomeningitis within a syngeneic or semiallogeneic system. Nature 1974; 248(5450): 701–2.

28. Lake P, Mitchison NA. Regulatory mechanisms in the immune response to cell-surface antigens. Cold Spring Harb Symp Quant Biol 1977; 41 Pt 2: 589–95.
29. Selin LK, Cornberg M, Brehm MA, Kim SK, Calcagno C, Ghersi D, et al. CD8 memory T cells: cross-reactivity and heterologous immunity. Semin Immunol 2004; 16(5): 335–47.
30. Pearson YE, Cheng Y, Selin LK, Puzone R, Celada F. Systematic simulation of cross-reactivity predicts ambiguity in Tk memory: it may save lives of the infected, but limits specificities vital for further responses. Autoimmunity 2011; 44(4): 315–27.
31. Cheng Y, Ghersi D, Calcagno C, Selin LK, Puzone R, Celada F. A discrete computer model of the immune system reveals competitive interactions between the humoral and cellular branch and between cross-reacting memory and naive responses. Vaccine 2009; 27(6): 833–45.
32. Fazekas de St Groth S, Webster RG. Disquisitions of Original Antigenic Sin. I. Evidence in man. J Exp Med 1966; 124(3): 331–45.
33. Monsalvo AC, Batalle JP, Lopez MF, Krause JC, Klemenc J, Hernandez JZ, et al. Severe pandemic 2009 H1N1 influenza disease due to pathogenic immune complexes. Nat Med 2011; 17(2): 195–9.
34. Rotman MB, Celada F. Antibody-mediated activation of a defective beta-D-galactosidase extracted from an Escherichia coli mutant. Proc Natl Acad Sci USA 1968; 60(2): 660–7.
35. Celada F, Strom R. Antibody-induced conformational changes in proteins. Q Rev Biophys 1972; 5(3): 395–425.
36. Celada F, Seiden PE. Teaching immunology: a Montessori approach using a computer model of the immune system. In: Celada F, Pernis B, editors. T Lymphocytes: Structure, Functions, Choices. New York: Plenum Press; 1992, pages 215–25.
37. Minsky M. A Framework for Representing Knowledge. MIT-AI Laboratory Memo 306, June, 1974.
38. Minsky M. The Society of Mind. New York: Simon and Schuster; 1988.
39. Langton CG. Artificial Life: An Overview. Cambridge, MA: MIT Press; 1995.
40. Holland JH. Genetic algorithms. Scientific American 1992; 267: 66–72.
41. Goldberg DE. Genetic Algorithms in Search, Optimization, and Machine Learning. Reading, MA: Addison Wesley; 1989.
42. Mohler RR, Bruni C, Gandolfi A. A systems approach to immunology. Proc IEEE 1980; 68: 964–90.

43. Lumb JR. Lymphocyte differentiation, repertoire development and migration: The need for mathematical models. Computers & Mathematics With Applications 1987; 14: 657–97.
44. Perelson AS. Theoretical Immunology, Part One. Redwood City, CA: Addison Wesley; 1988.
45. Perelson AS. Theoretical Immunology, Part Two. Redwood City, CA: Addison Wesley; 1988.
46. Perelson AS, Weisbuch G. Immunology for physicists. Rev Mod Phys 1997; 69: 1219–67.
47. Hege JS, Cole LJ. A mathematical model relating circulating antibody and antibody forming cells. J Immunol 1966; 97(1): 34–40.
48. Jilek M. The number of immunologically activated cells after repeated immunization. (A mathematical model). Folia Microbiol (Praha) 1971; 16(1): 12–23.
49. Bell GI. Mathematical model of clonal selection and antibody production. 3. TThe cellular basis of immunological paralysis. J Theor Biol 1971; 33(2): 378–98.
50. Bell GI. Predator-prey equations simulating an immune response. Math Biosci 1973; 16: 291–314.
51. Lotka AJ. Elements of Physical Biology. Baltimore: Williams & Wilkins Co.; 1925.
52. Volterra V. Variazioni e fluttuazioni del numero d'individui in specie animali conviventi. [Variations and fluctuations in the number of individuals in animal species living together.] Mem R Accad Naz dei Lincei 1926; 2: 31–113.
53. Bruni C, Giovenco MA, Koch G, Strom R. A dynamical model of humoral immune response. Math Biosci 1975; 27: 191–211.
54. Asachenkov AL, Belykh LH. Investigation of a mathematical model of viral disease. In: Marchuk GI, Nisevich NI, editors. Mathematical Methods in Clinical Practice [Matematiceskie metody v kliniceskoj praktike]. Novosibirsk: Nauka; 1978, pages 19–26.
55. Richter PH. A network theory of the immune system. Eur J Immunol 1975; 5(5): 350–4.
56. Hoffmann GW. A theory of regulation and self-nonself discrimination in an immune network. Eur J Immunol 1975; 5(9): 638–47.
57. Waltman P, Butz E. A threshold model of antigen-antibody dynamics. J Theor Biol 1977; 65(3): 499–512.
58. DeLisi C, Rescigno A. Immune surveillance and neoplasia. I. A minimal mathematical model. Bull Math Biol 1977; 39(2): 201–21.
59. Dibrov BF, Livshits MA, Volkenstein MV. Mathematical model of immune processes. J Theor Biol 1977; 65(4): 609–31.
60. Stewart J, Varela FJ. Morphogenesis in shape-space. Elementary meta-dynamics in a model of the immune network. J Theor Biol 1991; 153(4): 477–98.

61. Segel LA, Perelson AS. Exploiting the diversity of time scales in the immune system: a B-cell antibody model. J Stat Phys 1991; 63: 1113–31.
62. Dong X, Foteinou PT, Calvano SE, Lowry SF, Androulakis IP. Agent-based modeling of endotoxin-induced acute inflammatory response in human blood leukocytes. PLoS One 2010; 5(2): e9249.
63. De Oliveira S, De Oliveira P, Stauffer D. Evolution, Money, War, and Computers: Non-Traditional Applications of Computational Statistical Physics. Stuttgart-Leipzig: Vieweg+Teubner Verlag; 1999.
64. Ising E. Beitrag zur Theorie des Ferromagnetismus. [Contribution to the theory of ferromagnetism.] Zeitschrift für Physik 1925; 31: 253–8.
65. Kaufman M, Urbain J, Thomas R. Towards a logical analysis of the immune response. J Theor Biol 1985; 114(4):527–61.
66. Macal CM. Agent-based modeling and artificial life. In: Meyers RA, editor. Encyclopedia of Complexity and Systems Science. New York: Springer; 2009, pages 119–31.
67. Onsager L. Crystal statistics. I: a two-dimensional model with an order-disorder transition. Phys Rev 1944; 65: 117–49.
68. Laubenbacher RC, Jarrah AS, Mortveit HS, Ravi SS. A mathematical formalism for agent-based modeling. In: Meyers RA, editor. Encyclopedia of Complexity and Systems Science. New York: Springer; 2009, pages 160–76.
69. Marchuk GI. Mathematical modelling of immune response in infectious diseases. Dordrecht, Netherlands: Kluwer Academic Publishers; 1997.
70. Perelson AS, Oster GF. Theoretical studies of clonal selection: minimal antibody repertoire size and reliability of self-non-self discrimination. J Theor Biol 1979; 81(4): 645–70.
71. Farmer JD, Packard NH, Perelson AS. The immune system, adaptation and machine learning. Physica D 1986; 22: 187–204.
72. Zorzenon Dos Santos RM. Immune responses: getting close to experimental results with cellular automata models. In: Stauffer D, editor. Annual Reviews of Computational Physics, Vol. VI. Singapore: World Scientific; 1998, pages 159–202.
73. Zorzenon Dos Santos RM, Bernardes AT. Immunization and aging: a learning process in the immune network. Phys Rev Lett 1998; 81: 3034–7.
74. Forrest S, Javornik B, Smith RE, Perelson AS. Using genetic algorithms to explore pattern recognition in the immune system. Evolutionary Computation 1993; 1: 191–211.
75. Smith RE, Forrest S, Perelson AS. Searching for diverse, cooperative populations with genetic algorithms. Evolutionary Computation 1993; 1: 127–49.

76. Forrest S, Hofmeyr SA. Immunology as information processing. In: Segel LA, Cohen IR, editors. Design Principles for the Immune System and Other Distributed Autonomous Systems. New York: Oxford University Press; 2000, pages 361–87.
77. Dasgupta D. Artificial Immune Systems and Their Applications. Berlin Heidelberg: Springer-Verlag; 1999.
78. Weisbuch G, Atlan H. Control of the immune response. J Phys A 1988; 21: 189–92.
79. Dayan I, Havlin S, Stauffer D. Cellular automata generalisation of the Weisbuch-Atlan model for immune response. J Phys A 1988; 21: 2473–6.
80. Pandey R, Stauffer D. Immune response via interacting three-dimensional network of cellular automata. Journal de Physique 1989; 50: 1–10.
81. Pandey R, Stauffer D. Metastability with probabilistic cellular automata in an HIV infection. J Stat Phys 1990; 61: 235–40.
82. Succi S, Castiglione F, Bernaschi M. Collective dynamics in the immune system response. Phys Rev Lett 1997; 79: 4493–6.
83. Angermann BR, Klauschen F, Garcia AD, Prustel T, Zhang F, Germain RN, et al. Computational modeling of cellular signaling processes embedded into dynamic spatial contexts. Nat Methods 2012; 9(3): 283–9.
84. Folcik VA, An GC, Orosz CG. The Basic Immune Simulator: an agent-based model to study the interactions between innate and adaptive immunity. Theor Biol Med Model 2007; 4: 39.
85. Kalita JK, Chandrashekar K, Hans R, Selvam P. Computational modelling and simulation of the immune system. Int J Bioinform Res Appl 2006; 2(1): 63–88.
86. Duca K, Laubenbacher R, Polys NF, Luktuke R, Mcgee J, Shah J. The PathSim Project. http://www.vbi.vt.edu/~pathsim 2004.
87. Polys NF, Bowman DA, North C, Laubenbacher RC, Duca K. PathSim visualizer: an Information-Rich Virtual Environment framework for systems biology. Proceedings of the Ninth International Conference on 3D Web Technology 2004; 58: 7–14.
88. Efroni S, Harel D, Cohen IR. Toward rigorous comprehension of biological complexity: modeling, execution, and visualization of thymic T-cell maturation. Genome Res 2003; 13(11): 2485–97.
89. Bersini H, Klatzmann D, Six A, Thomas-Vaslin V. State-transition diagrams for biologists. PLoS One 2012; 7(7): e41165.
90. An G. Concepts for developing a collaborative *in silico* model of the acute inflammatory response using agent-based modeling. J Crit Care 2006; 21(1): 105–10.
91. Goldsby RA, Kindt TJ, Osborne BA. Kuby Immunology. 4th ed. New York: W.H. Freeman & Co.; 2000.

92. Murphy KP, Travers P, Walport M. Janeway's Immunobiology. 7th ed. New York: Garland Science; 2008.
93. Abbas AK, Lichtman AH. Basic Immunology: Functions and Disorders of the Immune System. 3rd ed. Philadelphia, PA: Saunders; 2011.
94. Castiglione F, Bernaschi M, Succi S. Simulating the immune response on a distributed parallel computer. Int J Mod Phys C 1997; 8(3): 527–45.
95. Seiden PE, Celada F. A model for simulating cognate recognition and response in the immune system. J Theor Biol 1992; 158(3): 329–57.
96. Morpurgo D, Serentha R, Seiden PE, Celada F. Modelling thymic functions in a cellular automaton. Int Immunol 1995; 7(4): 505–16.
97. Bezzi M, Celada F, Ruffo S, Seiden P. The transition between immune and disease states in a cellular automaton model of clonal immune response. Physica A 1997; 245: 145–63.
98. Lodish H, Baltimore D, Berk A, Zipursky SL, Matsudaira P, Darnell J. Molecular Cell Biology. New York: Scientific American Books, W.H. Freeman and Company; 1995.
99. Castiglione F, Santoni D, Rapin N. CTLs' repertoire shaping in the thymus: a Monte Carlo simulation. Autoimmunity 2011; 44(4): 261–70.
100. Salomaa A. Theory of Automata. Oxford: Pergamon Press; 1969.
101. Douek DC, Betts MR, Hill BJ, Little SJ, Lempicki R, Metcalf JA, et al. Evidence for increased T cell turnover and decreased thymic output in HIV infection. J Immunol 2001; 167(11): 6663–8.
102. Castiglione F, Sleitser V, Agur Z. Analyzing hypersensitivity to chemotherapy in a Cellular Automata model of the immune system. In: Preziosi L, editor. Cancer Modeling and Simulation. London: Chapman & Hall/CRC Press (UK); 2003, pages 333–65.
103. Castiglione F, Ribba B, Brass O. Comparing *in silico* results to *in vivo* and *ex vivo* of influenza-specifci immune responses after vaccination or infection in humans. In: Baschieri S, editor. Innovation in Vaccinology: from design, through to delivery and testing. XII ed. New York: Springer; 2012, pages 17–43.
104. Uhlenbeck GE, Ornstein LS. On the theory of Brownian motion. Phys Rev 1930; 36: 823–41.
105. Vacanti MP, Roy A, Cortiella J, Bonassar L, Vacanti CA. Identification and initial characterization of spore-like cells in adult mammals. J Cell Biochem 2001; 80(3): 455–60.
106. Shay JW, Wright WE. Hayflick, his limit, and cellular ageing. Nat Rev Mol Cell Biol 2000; 1(1): 72–6.
107. Slifka MK, Antia R, Whitmire JK, Ahmed R. Humoral immunity due to long-lived plasma cells. Immunity 1998; 8(3): 363–72.

108. Celada F. Quantitative studies of the adoptive immunological memory in mice. II. Linear transmission of cellular memory. J Exp Med 1967; 125(2): 199–211.
109. Celada F. The cellular basis of immunologic memory. Prog Allergy 1971; 15: 223–67.
110. Weinand RG. Somatic mutation, affinity maturation and the antibody repertoire: a computer model. J Theor Biol 1990; 143(3): 343–82.
111. Castiglione F, Bernaschi M. HIV-1 strategies of immune evasion. Int J Mod Phys C 2006; 16: 1869–79.
112. Castiglione F, Bernaschi M, Succi S. A high performance simulator of the immune response. Future Gen Comput Syst 1999; 15: 333–42.
113. Aho AV, Hopcroft JE, Ullman JD. Data Structures and Algorithms. Boston, MA: Addison-Wesley Longman Publishing Co.; 1983.
114. Bousso P, Robey E. Dynamics of CD8+ T cell priming by dendritic cells in intact lymph nodes. Nat Immunol 2003; 4(6): 579–85.
115. Miller MJ, Wei SH, Parker I, Cahalan MD. Two-photon imaging of lymphocyte motility and antigen response in intact lymph node. Science 2002; 296(5574): 1869–73.
116. Miller MJ, Wei SH, Cahalan MD, Parker I. Autonomous T cell trafficking examined *in vivo* with intravital two-photon microscopy. Proc Natl Acad Sci USA 2003; 100(5): 2604–9.
117. Francis K, Palsson BO. Effective intercellular communication distances are determined by the relative time constants for cyto/chemokine secretion and diffusion. Proc Natl Acad Sci USA 1997; 94(23): 12258–62.
118. Baldazzi V, Paci P, Bernaschi M, Castiglione F. Modeling lymphocyte homing and encounters in lymph nodes. BMC Bioinformatics 2009; 10: 387.
119. Segovia-Juarez JL, Ganguli S, Kirschner D. Identifying control mechanisms of granuloma formation during M. tuberculosis infection using an agent-based model. J Theor Biol 2004; 231(3): 357–76.
120. Lee GR, Wintrobe MM. Wintrobe's Clinical Hematology. 9th ed. Philadelphia: Lea & Febiger; 1993.
121. Del Prete G, Maggi E, Parronchi P, Chretien I, Tiri A, Macchia D, et al. IL-4 is an essential factor for the IgE synthesis induced *in vitro* by human T cell clones and their supernatants. J Immunol 1988; 140(12): 4193–8.
122. Bernaschi M, Succi S, Castiglione F. Large-scale cellular automata simulations of the immune system response. Phys Rev E Stat Phys Plasmas Fluids Relat Interdiscip Topics 2000; 61(2): 1851–4.
123. Bernaschi M, Castiglione F. Design and implementation of an immune system simulator. Comput Biol Med 2001; 31(5): 303–31.

124. Kohler B, Puzone R, Seiden PE, Celada F. A systematic approach to vaccine complexity using an automaton model of the cellular and humoral immune system. I. Viral characteristics and polarized responses. Vaccine 2000; 19(7-8): 862–76.
125. Shoenfeld Y. The idiotypic network in autoimmunity: antibodies that bind antibodies that bind antibodies. Nat Med 2004; 10(1): 17–8.
126. Holgate ST. Science, medicine, and the future. Allergic disorders. BMJ 2000; 320(7229): 231–4.
127. Santoni D, Pedicini M, Castiglione F. Implementation of a regulatory gene network to simulate the TH1/2 differentiation in an agent-based model of hypersensitivity reactions. Bioinformatics 2008; 24(11): 1374–80.
128. Mendoza L. A network model for the control of the differentiation process in Th cells. Biosystems 2006; 84(2): 101–14.
129. McCune JM. The dynamics of CD4+ T-cell depletion in HIV disease. Nature 2001; 410(6831): 974–9.
130. Nelson PW, Perelson AS. Mathematical analysis of HIV-1. Dynamics in vivo. SIAM Rev 1999; 41: 3–44.
131. Finn OJ. Cancer vaccines: between the idea and the reality. Nat Rev Immunol 2003; 3(8): 630–41.
132. Nouri-Shirazi M, Banchereau J, Fay J, Palucka K. Dendritic cell based tumor vaccines. Immunol Lett 2000; 74(1): 5–10.
133. Stevenson FK, Rice J, Zhu D. Tumor vaccines. Adv Immunol 2004; 82: 49–103.
134. Schneider J, Gilbert SC, Blanchard TJ, Hanke T, Robson KJ, Hannan CM, et al. Enhanced immunogenicity for CD8+ T cell induction and complete protective efficacy of malaria DNA vaccination by boosting with modified vaccinia virus Ankara. Nat Med 1998; 4(4): 397–402.
135. Yewdell JW, Bennink JR. Immunodominance in major histocompatibility complex class I-restricted T lymphocyte responses. Annu Rev Immunol 1999; 17: 51–88.
136. Castiglione F, Toschi F, Bernaschi M, Succi S, Benedetti R, Falini B, et al. Computational modeling of the immune response to tumor antigens. J Theor Biol 2005; 237(4): 390–400.
137. Etlinger HM, Gillessen D, Lahm HW, Matile H, Schonfeld HJ, Trzeciak A. Use of prior vaccinations for the development of new vaccines. Science 1990; 249(4967): 423–5.
138. Rice J, Buchan S, Stevenson FK. Critical components of a DNA fusion vaccine able to induce protective cytotoxic T cells against a single epitope of a tumor antigen. J Immunol 2002; 169(7): 3908–13.
139. Palmowski MJ, Choi EM, Hermans IF, Gilbert SC, Chen JL, Gileadi U, et al. Competition between CTL narrows the immune response induced by prime-boost vaccination protocols. J Immunol 2002; 168(9): 4391–8.

140. Rapin N, Lund O, Bernaschi M, Castiglione F. Computational immunology meets bioinformatics: the use of prediction tools for molecular binding in the simulation of the immune system. PLoS One 2010; 5(4):e9862.
141. Korber B, LaBute M, Yusim K. Immunoinformatics comes of age. PLoS Comp Biol 2006; 2: 484–92.
142. Petrovsky N, Brusic V. Computational immunology: The coming of age. Immunol Cell Biol 2002; 80(3): 248–54.
143. Petrovsky N, Brusic V. Bioinformatics for study of autoimmunity. Autoimmunity 2006; 39(8): 635–43.
144. Darve E, Pohorille A. Calculating free energies using average force. J Chem Phys 2001; 115: 9169–83.
145. Lundegaard C, Lund O, Kesmir C, Brunak S, Nielsen M. Modeling the adaptive immune system: predictions and simulations. Bioinformatics 2007; 23(24): 3265–75.
146. Sette A, Fleri W, Peters B, Sathiamurthy M, Bui HH, Wilson S. A roadmap for the immunomics of category A-C pathogens. Immunity 2005; 22(2): 155–61.
147. Lin HH, Ray S, Tongchusak S, Reinherz EL, Brusic V. Evaluation of MHC class I peptide binding prediction servers: applications for vaccine research. BMC Immunol 2008; 9: 8.
148. Lin HH, Zhang GL, Tongchusak S, Reinherz EL, Brusic V. Evaluation of MHC-II peptide binding prediction servers: applications for vaccine research. BMC Bioinformatics 2008; 9 Suppl 12: S22.
149. Yewdell JW, Reits E, Neefjes J. Making sense of mass destruction: quantitating MHC class I antigen presentation. Nat Rev Immunol 2003; 3(12): 952–61.
150. Lund O, Nielsen M, Lundegaard C, Kesmir C, Brunak S. Immunological Bioinformatics. Cambridge, MA: MIT Press; 2005.
151. Parker JM, Guo D, Hodges RS. New hydrophilicity scale derived from high-performance liquid chromatography peptide retention data: correlation of predicted surface residues with antigenicity and X-ray-derived accessible sites. Biochemistry 1986; 25(19): 5425–32.
152. Levitt M, Greer J. Automatic identification of secondary structure in globular proteins. J Mol Biol 1977; 114(2): 181–239.
153. Miyazawa S, Jernigan RL. Residue-residue potentials with a favorable contact pair term and an unfavorable high packing density term, for simulation and threading. J Mol Biol 1996; 256(3): 623–44.
154. Miyazawa S, Jernigan RL. An empirical energy potential with a reference state for protein fold and sequence recognition. Proteins 1999; 36(3): 357–69.
155. Miyazawa S, Jernigan RL. Identifying sequence-structure pairs undetected by sequence alignments. Protein Eng 2000; 13(7): 459–75.

156. Lund O, Nielsen M, Kesmir C, Petersen AG, Lundegaard C, Worning P, et al. Definition of supertypes for HLA molecules using clustering of specificity matrices. Immunogenetics 2004; 55(12): 797–810.
157. Nielsen M, Lundegaard C, Worning P, Hvid CS, Lamberth K, Buus S, et al. Improved prediction of MHC class I and class II epitopes using a novel Gibbs sampling approach. Bioinformatics 2004; 20(9): 1388–97.
158. Nielsen M, Lundegaard C, Blicher T, Lamberth K, Harndahl M, Justesen S, et al. NetMHCpan, a method for quantitative predictions of peptide binding to any HLA-A and -B locus protein of known sequence. PLoS One 2007; 2(8): e796.
159. Nielsen M, Lundegaard C, Lund O. Prediction of MHC class II binding affinity using SMM-align, a novel stabilization matrix alignment method. BMC Bioinformatics 2007; 8: 238.
160. Nielsen M, Lundegaard C, Worning P, Lauemoller SL, Lamberth K, Buus S, et al. Reliable prediction of T-cell epitopes using neural networks with novel sequence representations. Protein Sci 2003; 12(5): 1007–17.
161. Buus S, Lauemoller SL, Worning P, Kesmir C, Frimurer T, Corbet S, et al. Sensitive quantitative predictions of peptide-MHC binding by a 'Query by Committee' artificial neural network approach. Tissue Antigens 2003; 62(5): 378–84.
162. Kumar N, Hendriks BS, Janes KA, de Graaf D, Lauffenburger DA. Applying computational modeling to drug discovery and development. Drug Discov Today 2006; 11(17-18): 806–11.
163. Davies MN, Flower DR. Harnessing bioinformatics to discover new vaccines. Drug Discov Today 2007; 12(9-10): 389–95.
164. Woelke AL, von Eichborn J, Murgueitio MS, Worth CL, Castiglione F, Preissner R. Development of immune-specific interaction potentials and their application in the multi-agent-system VaccImm. PLoS One 2011; 6(8): e23257.

Subject Index

S

T

V

Z

Index of Persons Named

Colour Plate Section

Chapter 1

Figure 2. The lamprey captured near the coast of New Jersey (left) and its unusual mouth (right). Reproduction with permission from the author Doug Cutler.

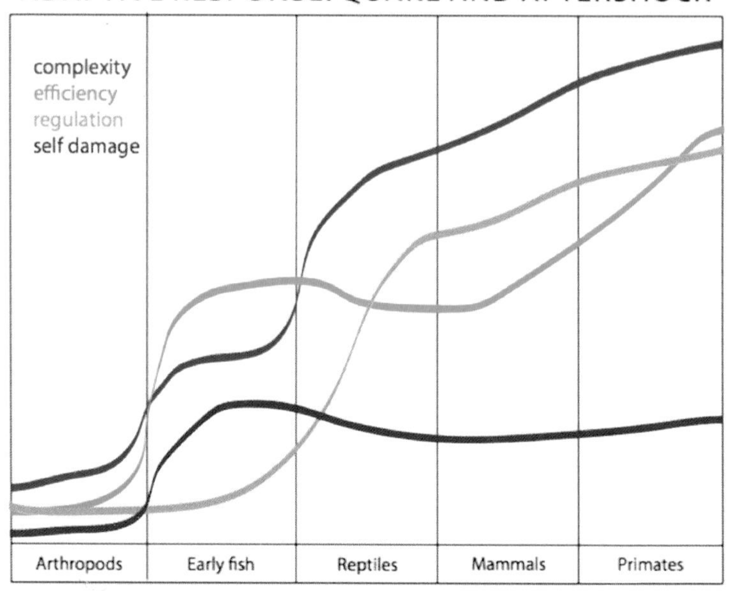

Figure 4. (Artist: Valentina Celada) A 'paleohistory' of the immune system: arthropods and molluscs of today exhibit the oldest kind of defence, non-specific, cellular and humoral that dates back 800 to 900 million years, and has allowed multicellular life to populate the primitive *bouillon de culture*. About 500 to 600 million years ago, *early fish* form tiny vertebrae and are surprised by one of the major revolutions in biology; thanks to lymphoid cells, the response becomes specific and the diversity grows at all levels and in repertoires. Responses are effective, but friendly fire (auto-aggression) is staggering. Reptiles, 300 million years ago, and then the dinosaurs, inherit a faulty weapon that shoots also backward. An immunological cause for the extinction of the latter is difficult to prove, but even more to disprove. Mammals, at 100 million years ago, build up the system in the only evolutionarily correct way: every cell, every clone stimulated is controlled and regulated (killed) if it shows auto-aggression. The immune system becomes multi-organ, allowing for special rooms for cells to meet and kill the suspected. And primates? Well, primates, appearing about 60 million years ago, must have been engaged in the last step of immune evolution, which is fully active even today. Memory cells have become central in the defence, because they are enjoying large numbers and ready helper cooperation. However memory thwarts new specific responses and this is against the trend that began in the lamprey. The cross-reactive, low-affinity dominance is arguably the cause of senescent immunity, whose principal target is man (at 25 million years ago), and especially modern man whose life has dramatically increased 100 years ago (see section 1.50).

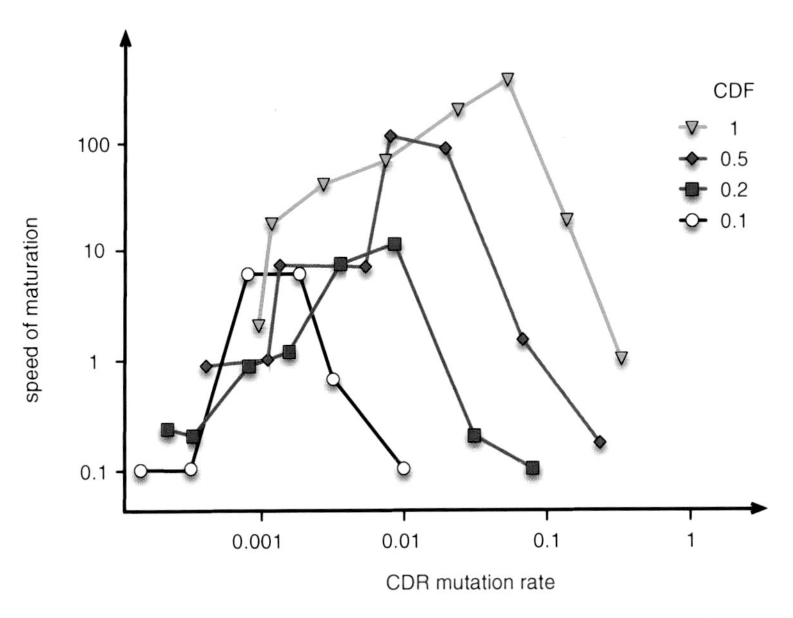

Figure 6. This graph shows an experiment *in silico* that illustrates the positive-negative effect on affinity maturation observed when the mutation rate increases. The experiment is run in IMMSIM; in the abscissa is the complementarity-determining region (CDR) mutation rate, growing over three orders of magnitude (from 10^{-3} to 1 event per time unit), in the ordinate is the speed of maturation of the affinity. The plot combines four similar experiments differing for the fraction of mutations in the chromosome that actually affects the CDR (This is called the CDR fraction or CDF) (abridged from [7]).

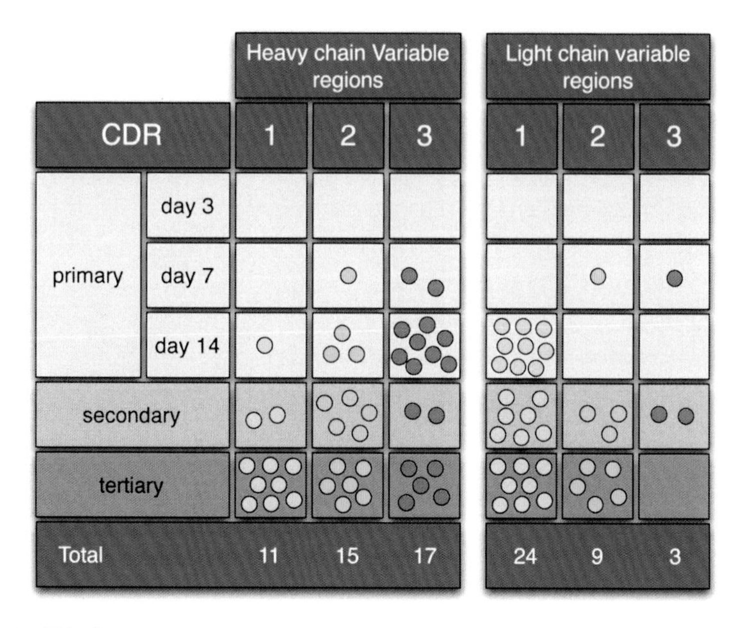

Figure 7. This figure shows the classic observation by Cesar Milstein that mutations appear in the CDR of the antibody, forming clones in late primary response and at higher rates during the secondary and tertiary responses. All three time frames are characterised by active cell division activity, especially the latter two. For this experiment, Milstein has used the technique of hybridoma that he has developed to produce monoclonal antibodies: antigen-challenged B-cells are hybridised with myeloma cells; immortalised cells are selected for antibody production, then grown *in vitro* and cloned from single cells and studied [8].

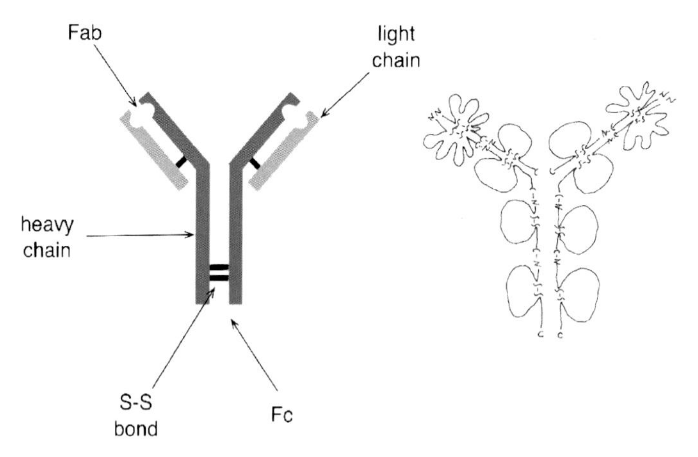

Figure 8. The left panel shows the structure of the antibody, as it was established around 1960 [14, 15]. It shows a molecule with a sedimentation coefficient equal to 7S consisting of two identical dimers, each formed of one heavy and one light chain. The right panel shows the way the antibody is an assemblage with a total of 16 strings, each consisting of around one hundred amino acids.

Figure 11. Oak Ridge National Laboratories, the Biology Division, in a photo of 1976, kindly provided by the Oak Ridge ORNL Library. The building, identified as Y12, had been planned as part of the Manhattan Project in the 1940s, and should have hosted an industrial scale ion-spectrograph for the purpose of isolating U335, the fuel of atomic bombs. When another method of Uranium purification proved successful, the works at Y12 were discontinued, and the site was occupied by the ORNL Biology Division in the mid-1950s. Some preliminary tests with Uranium must have already been made, since during the first years the biologists were not allowed to access some parts of the large building on account of the radioactive contamination level.

Figure 13. The group talking with Umberto Eco: (from left to right) Umberto Eco, Eli Sercarz, Franco Celada and Masaru Taniguchi. Photo courtesy of Rabyn Blake.

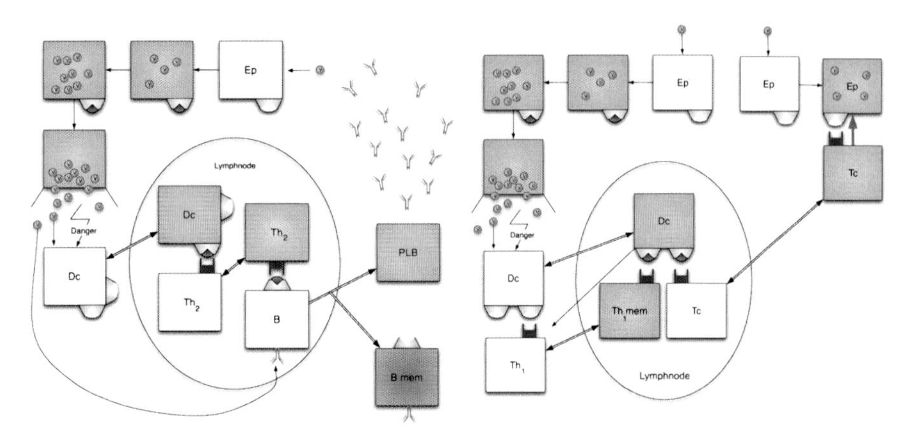

Figure 15. *Left panel*: The humoral response. Starting from the top and proceeding counter clockwise, the virus enters an epithelial cell Ep and begins to multiply. In succession, Ep displays part of the virus (V) on its MHC class-1 then lyses, liberating virus particles. Some of these are picked up by a resting macrophage or dendritic cell (Dc), which also perceives a damage or danger signal. Thus, it transits to an aroused state (light blue shaded Dc), and moves towards a regional lymph node. Here it displays a viral peptide on its class-1 MHC molecule, and if a virgin Th2 makes specific contact, it is activated and will be able to, in turn, activate a B-cell. The latter multiplies and differentiates into antibody-secreting plasma cells (PLB) and B memory cells (B mem). The antibodies are able to inactivate live viruses but are harmless for virus nested inside target cells.

Right panel: The cellular response. The triggering is identical to the humoral up to the activation of the APC, which then binds, on its receptors, both an effector and a helper T-cell. Both are activated by the Dc, and the 'help' reaches the effector (Tc) thanks to the proximity of the two T-cells. Consequently the effector multiplies and its offspring are ready to bind the class-1 receptors of any infected epithelial cell and kill it, thereby destroying also the incomplete virus particles present in the assembly line. There is consensus that the 'help' can still reach the effector even in case of a temporal discrepancy at the time of the binding of the T-cells to the APC.

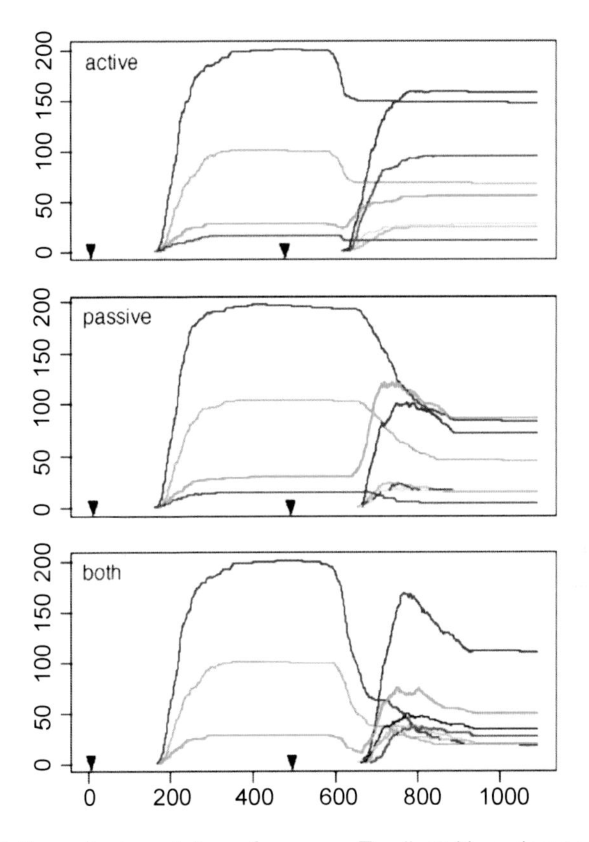

Figure 16. Mathematical modeling of memory T-cell attrition, demonstrating active, passive, or both mechanisms: Tc memory clone dynamics during the primary and secondary response to cross-reacting viruses, under three scenarios of attrition simulated by IMMSIM. Abscissa: time steps. Ordinate: Tc memory clone cell counts. Arrows at time step 1 and 500 indicate the time of inoculum of V1 (70 particles) and V2 (120 particles), respectively. V1 clearance is complete at time step 200 and V2 clearance at time step 700. Reprinted with permission from [29].

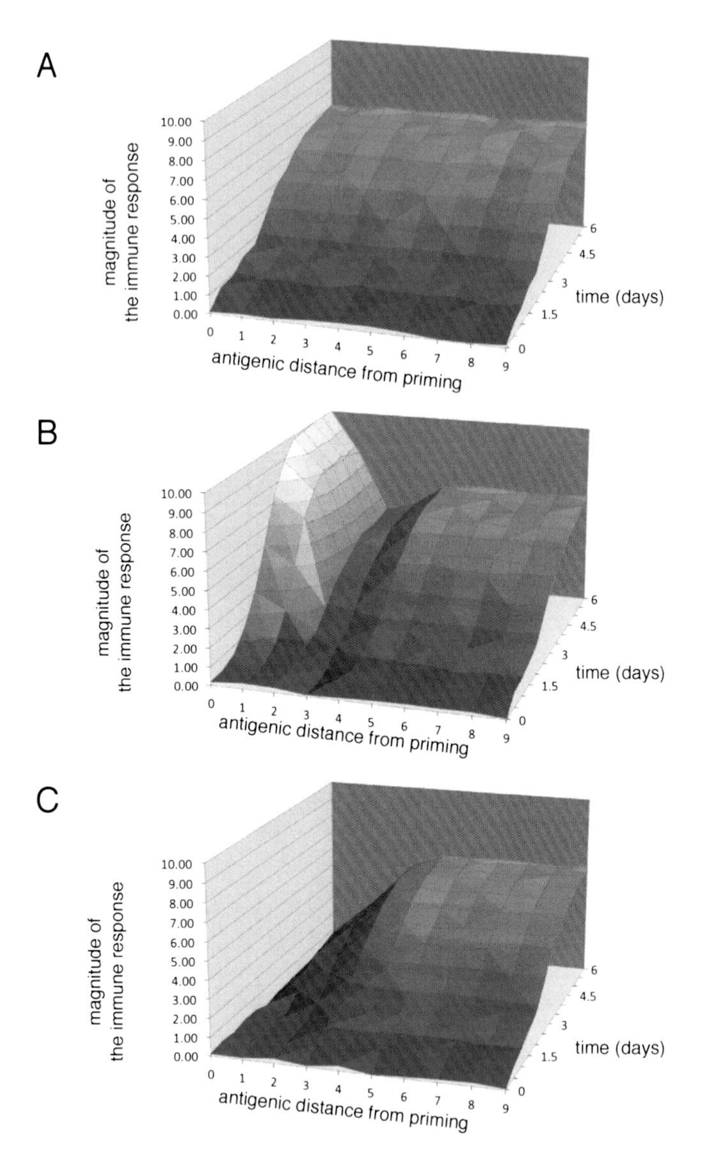

Figure 17. Experiment *in silico* demonstrating the MaN (Memory anti Naïve) phenomenon [31]. In panel A, naïve T-cells develop a primary response against ten virtual viruses differing from each other by one bit (out of 16) in antigen specificity: results of the ten runs, labeled with the antigen distance from 0 to 9, are shown side by side in the 3D representation. Each curve shows the kinetics of the cytotoxic response. Thanks to the large repertoire available, the naïve responses to the different viruses are not significantly different. In panel B, the cell population contains not only naïve cells but also memory cells primed with virus '0'. The 3D pattern shows the secondary

Figure 17. contd....

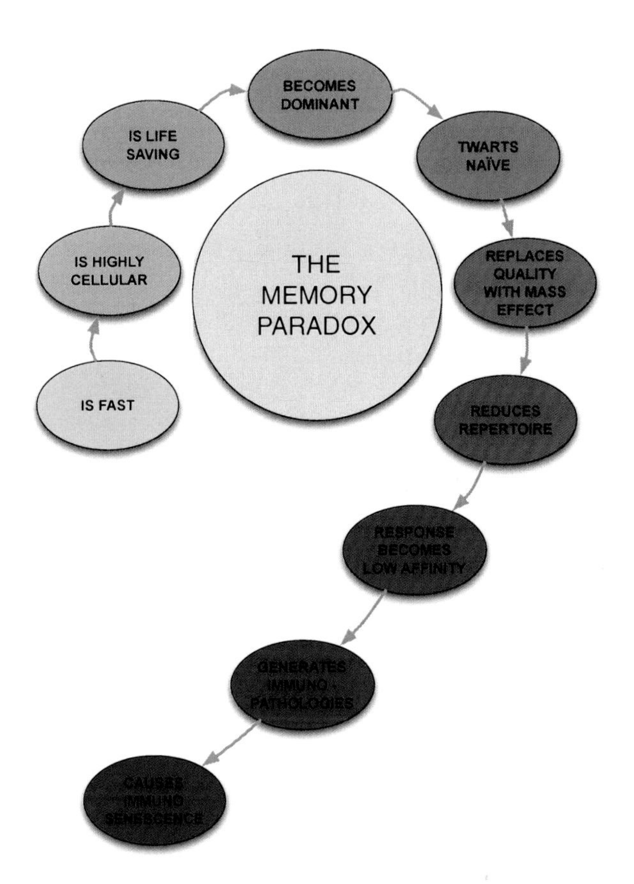

Figure 18. The attributes of immunological memory in colour-coded balloons according to positive (green) and negative (red) effects on the immune system.

Figure 17. contd.

response developing in column 0 and the decreasing cross-reaction in columns 1, 2 and 3. Note the higher speed and magnitude of the memory response. The primary response is completely wiped out in column 1, and severely thwarted in columns 2 and 3. Only from column 4 on, is the primary response undisturbed. The MaN effect is enhanced by the memory clone size and speed of growth, which together outcompete the naïve cells. In panel C, the same data of the experiment in panel B are plotted, but the memory response is artificially cancelled to better illustrate the 'hole' caused by memory on the primary response.

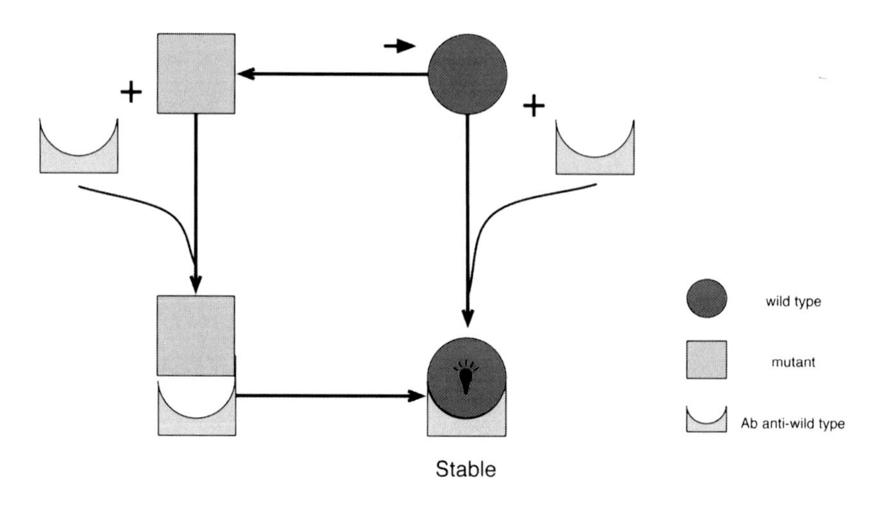

Figure 21. Equilibrium of a general antibody-mediated activation of a functional molecule. (see description in the text.)

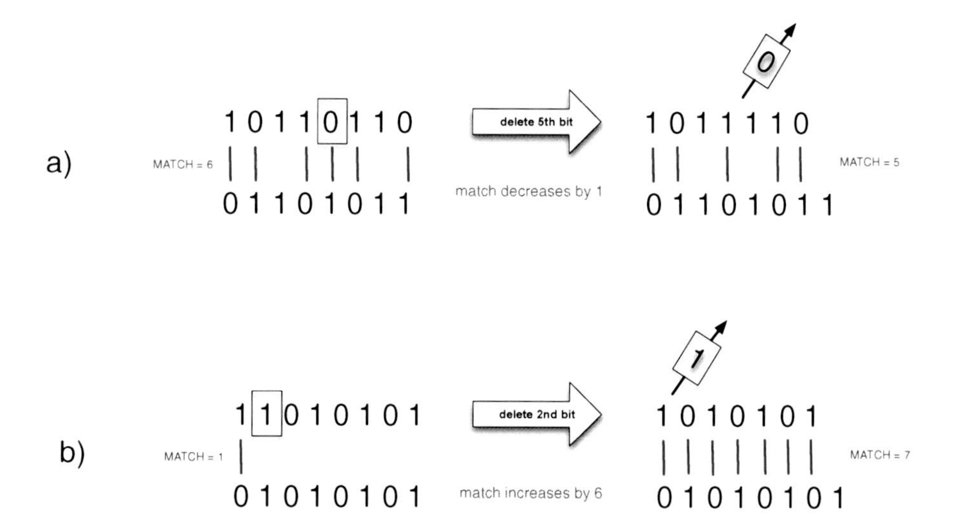

Figure 22. Binary strings can be used to identify objects and determine their mutual attraction (IMMSIM uses 24 bits for each paratope and each epitope, and for each receptor and each specific ligand). When these objects come in proximity, the respective strings align and each of their bits meets their opposite one. Attraction is only exerted between 1 and 0 and 0 and 1 (bit match). The sum of matches determines the affinity of the interaction. In the scheme, one vertical red bar symbolises a match.

Chapter 3

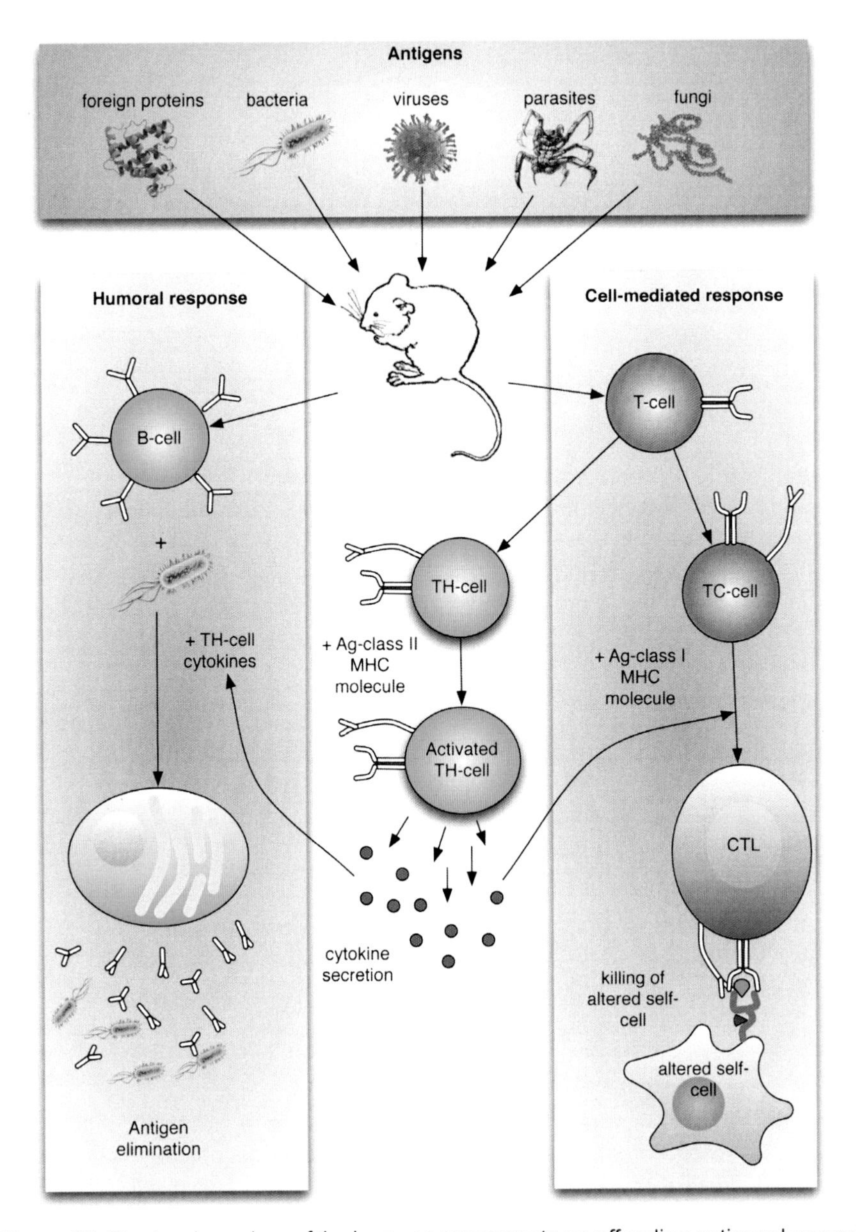

Figure 23. The two branches of the immune response to an offending antigen: humoral response, mediated by the production of antibodies, and the cellular response, mediated by the action of activated cytotoxic T lymphocytes. C-ImmSim implements both and enables the representation of various pathogens as virus and bacteria.

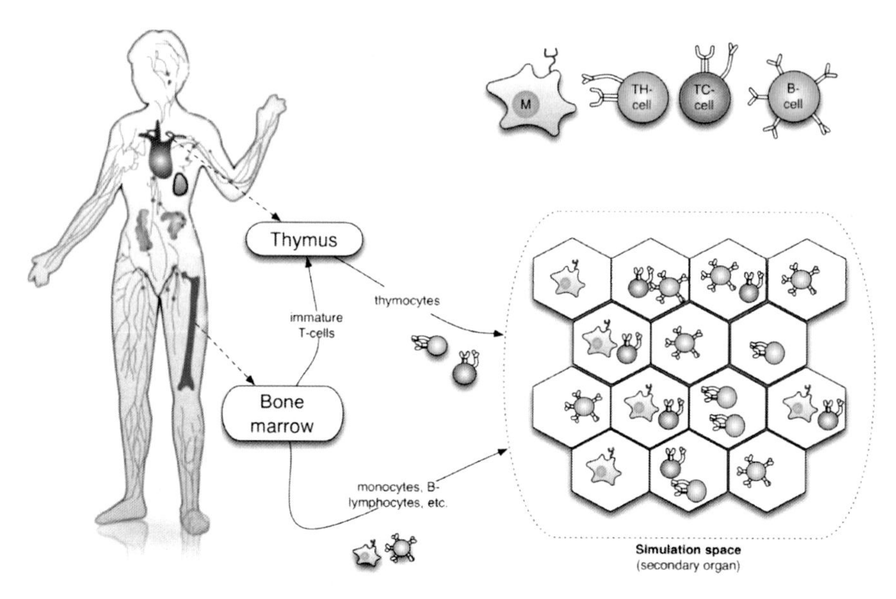

Figure 24. The three anatomical compartments modelled are the thymus, the bone marrow and a portion of a generic secondary organ (the picture of the human body is adapted from [98]).

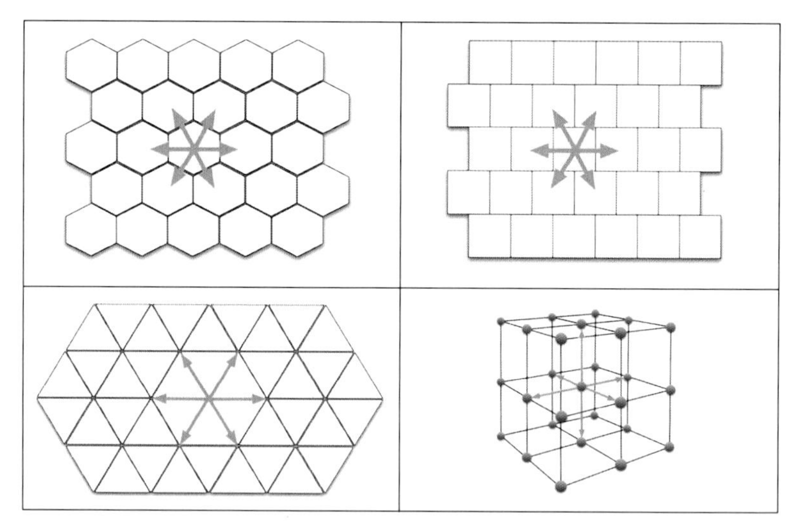

Figure 25. The space is discrete. In the bi-dimensional version of C-ImmSim the grid is a hexagonal lattice (top, left) or square-shifted (top, right). This is equivalent to the triangular lattice if you look at the edges instead of the nodes (bottom-left). In the three-dimensional version, the space is a Cartesian lattice (bottom-right).

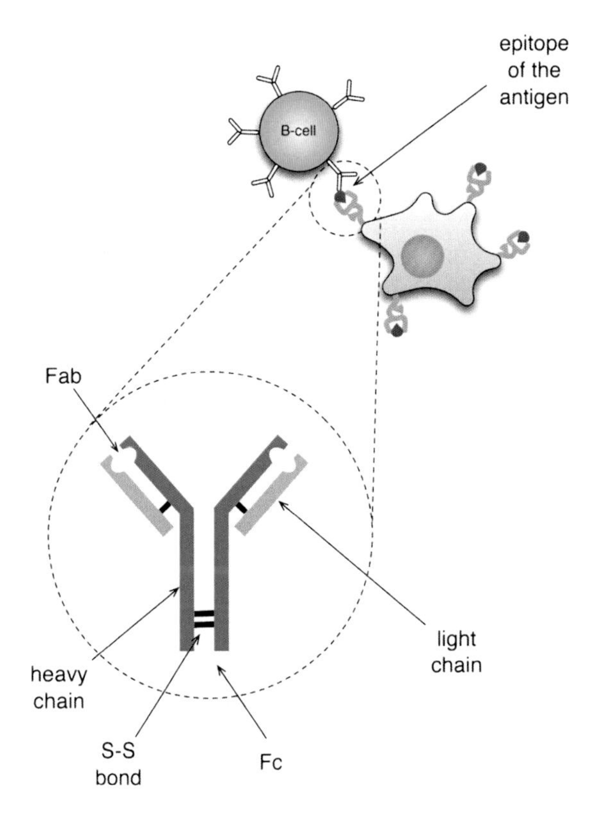

Figure 33. The major type of immunoglobulin, IgG, is the one that is often used to represent the antibodies in C-ImmSim and the only one used in IMMSIM while a subdivision in classes IgG1, IgG2 or IgE has been done to account for significant processes in specific diseases like hypersensitivity (section 5.1) or flu.

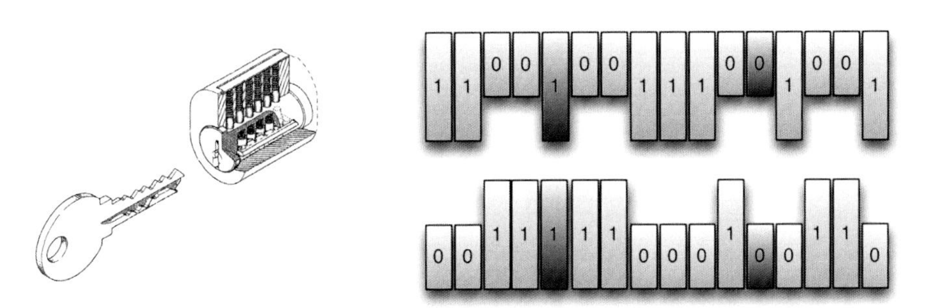

Figure 34. The match between two strings reminds one of that between a lock and a key. It is computed as the Hamming distance in the space of all possible binary strings of a certain length. In this example, 16 bits are used. The mismatch is two because two pins (those in red) do not pair.

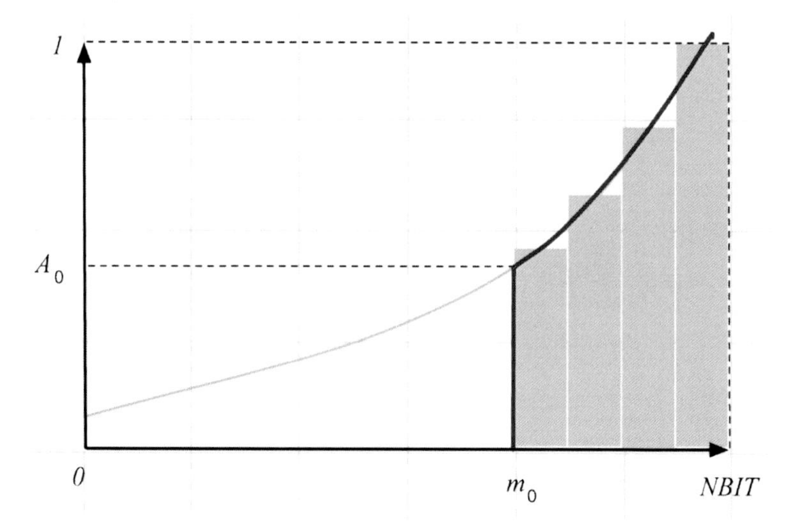

Figure 35. Affinity function. The probability to bind is a function of the Hamming distance between the two binding sites of the interacting entities. A match of m_0 or greater is required to have positive affinity.

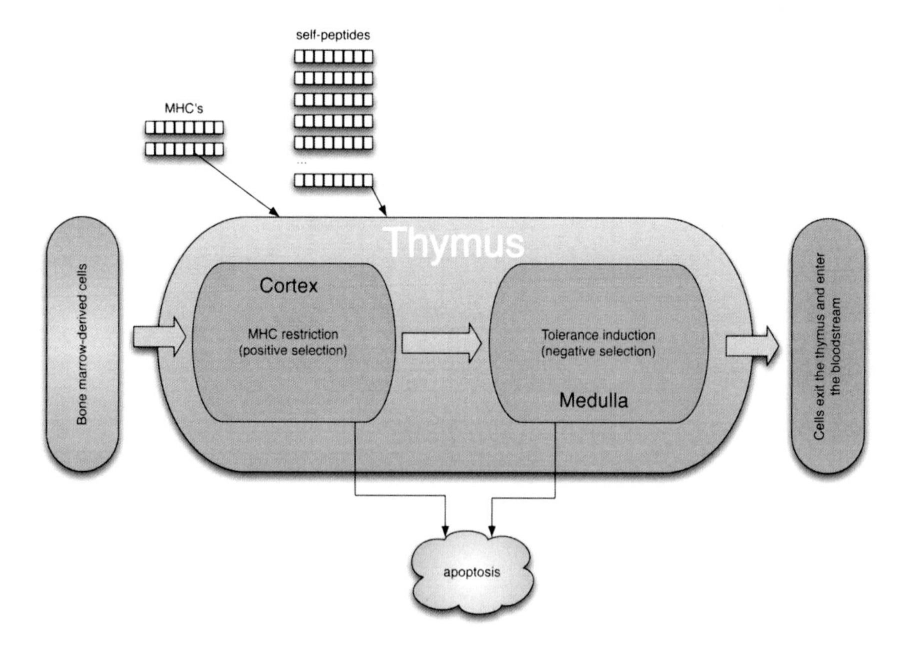

Figure 37. The two-layer filter realised by the thymus to eliminate auto-reactive T lymphocytes.

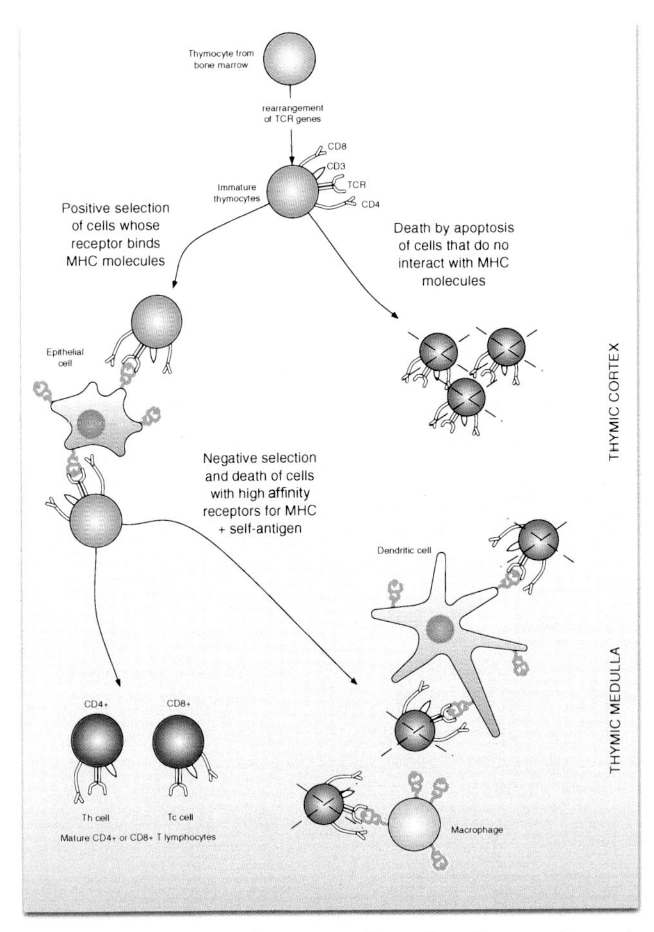

Figure 38. This cartoon shows the maturation of a thymocyte and subsequent involvement in the humoral and cellular immune response. In C-ImmSim, the maturation process is not explicitly modelled; just the activation from a resting state is necessary for a lymphocyte to enter into action.

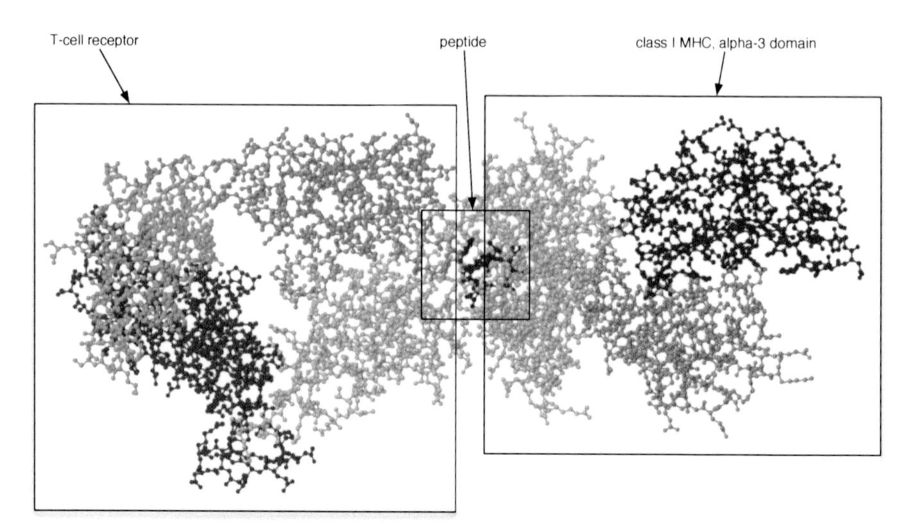

Figure 39. TCR binding is allowed in particular regions of the MHC/peptide. In the model approximation, receptors and peptides are binary strings. They stand for the effective binding regions only. Figure made with Jmol: an open-source Java viewer for chemical structures in 3D. http://www.Jmol.org.

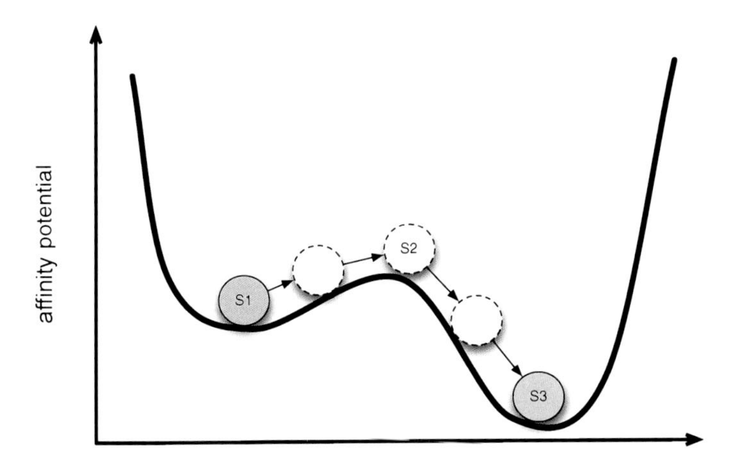

Figure 40. A metastable state of weaker recognition of the antigen (S1), in the initial adaptive immune response, corresponds to a local optima. It can shift through intermediate 'saddle' configurations (S2) to reach a stable state (S3) of stronger recognition corresponding to a better global affinity to the antigen.

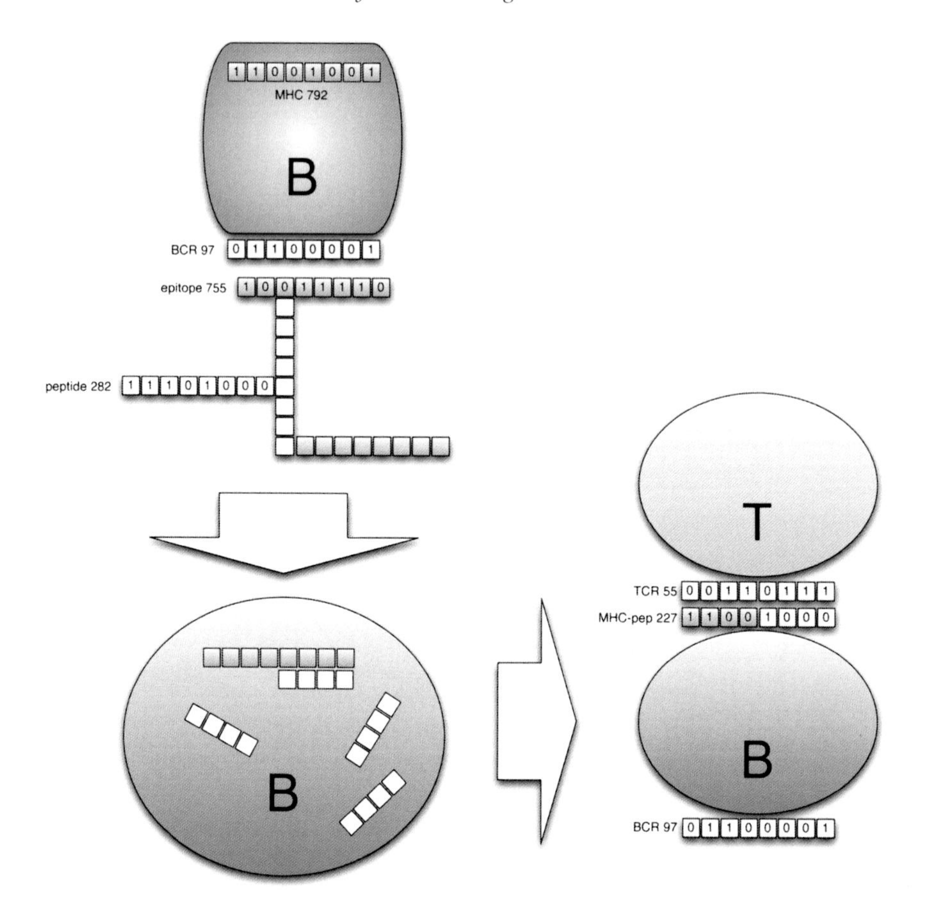

Figure 41. A B-cell binds the antigen. After the antigen digestion and presentation together with the MHC molecule on its surface, it interacts with an active Th cell upon matching with the T-cell receptor. This is the 'pivotal' cooperation step discussed in section 1.16.

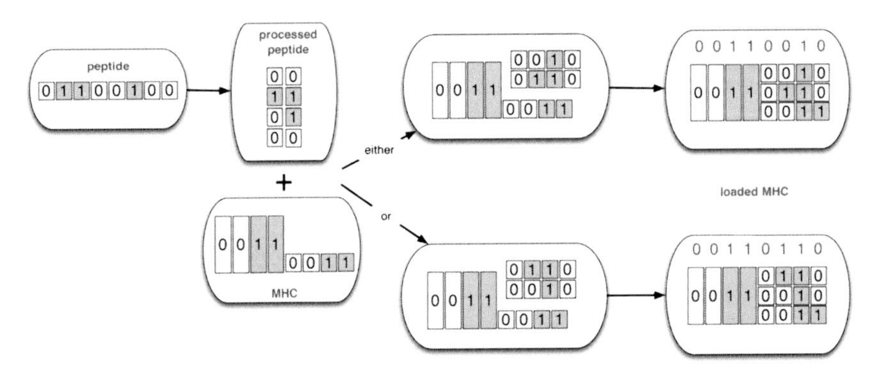

Figure 42. In the binary representation of the peptides, they are arranged in two substrings: one binds to the MHC and the other is shown in the MHC–peptide complex. In contrast, the MHC molecule offers just one binding site for the peptide, that is conventionally chosen to be the right half of the binary string. So, when the right half is bound to one half of the peptide (either left or right), the one that is shown in the MHC/peptide complex is the left part (adapted from [96]).

Chapter 4

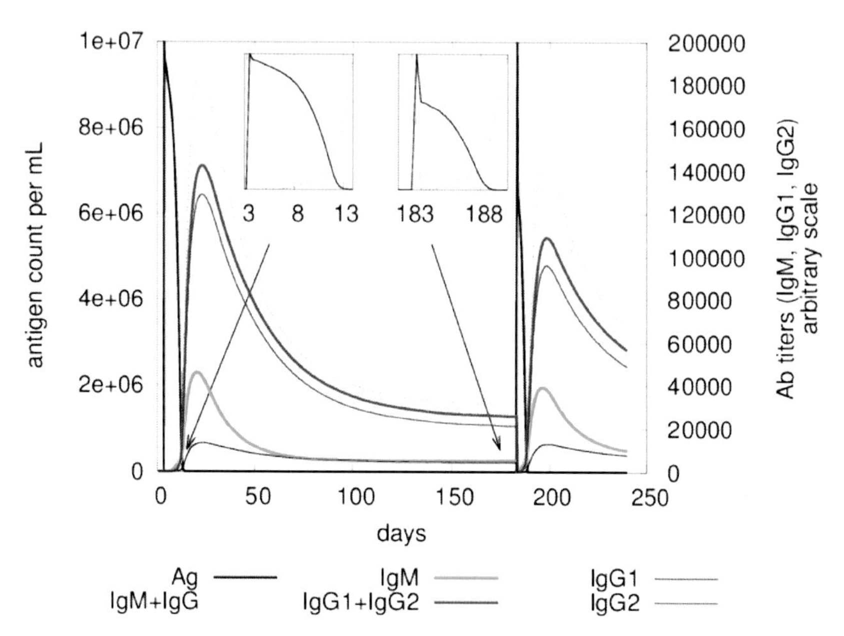

Figure 44. This plot shows the number of antigenic molecules in a millilitre of blood versus time in a vaccination experiment. Two injections are administered at different time steps to observe a faster secondary immune response due to the memory effect elicited from the priming injection. This plot shows, also, the total number of specific antibodies produced (on the y2 axis).

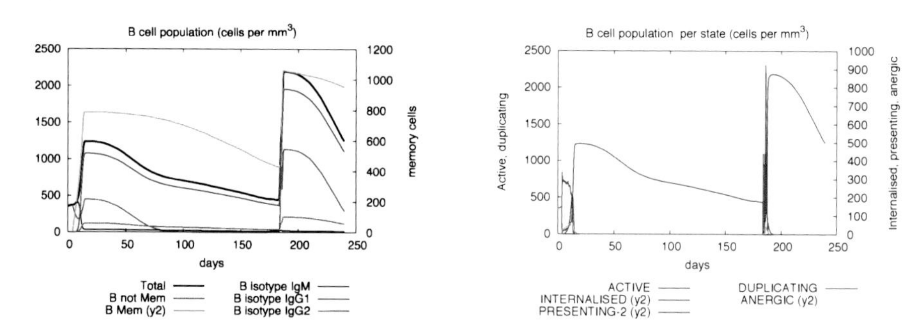

Figure 45. On the left is the number of B-cells in a cubic millimetre of lymph node tissue or (for convenience) in a micro litre of peripheral blood, together with the relative number of memory cells (on the y2 axis). On the right the same number is plotted, but for each possible internal state the B-cells can take.

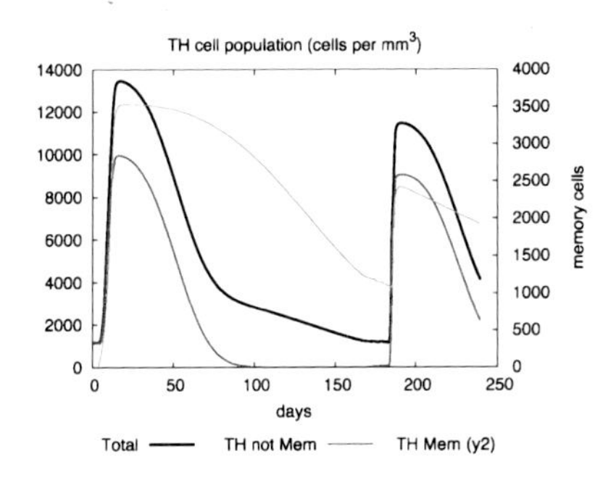

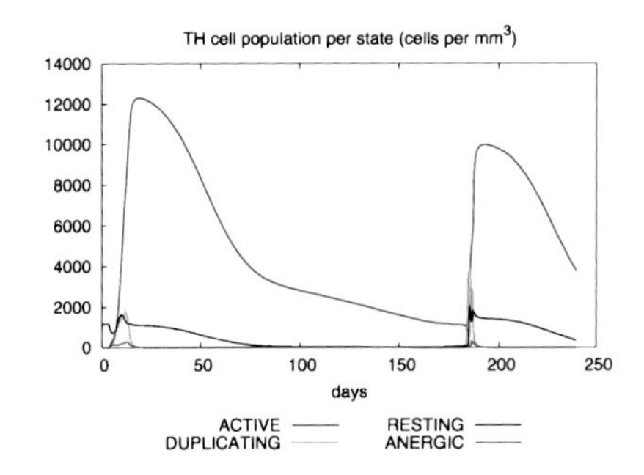

Figure 46. T helper lymphocytes during a vaccination experiment.

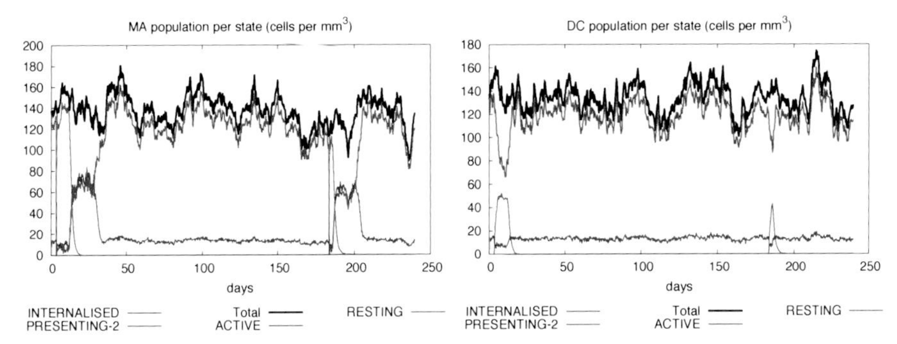

Figure 47. As antigen-processing cells, macrophages (MA, left panel) and dendritic cells (DC, right panel) can take many internal states like the intermediate state, INTERNALISED, preceding the state PRESENTING-2 and following the phagocytosis of the antigen. The INTERNALISED is an intermediate state that is quickly changed in a time step equivalent to 6–8 hours; this is the reason why the plot does not show any cell in this state.

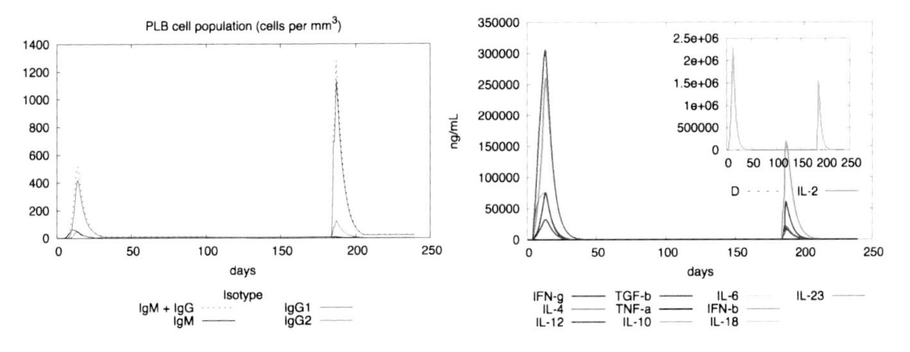

Figure 48. Plasma B-cells (PLB) peak after antigen stimulation. The predominant isotype is the IgG (left panel). In the right panel, the plot shows the concentration of various interleukins, each carrying a different message for the activated immune system.

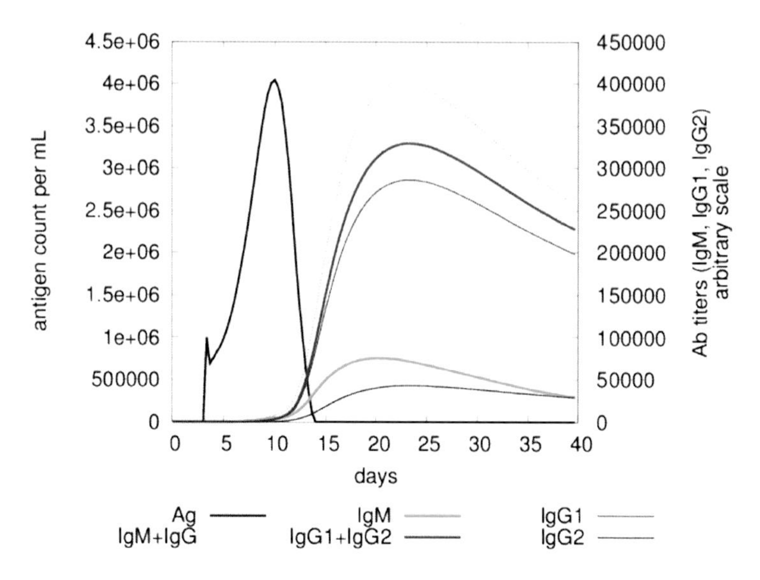

Figure 50. Bacteria replicate freely until an effective immune response is mounted about one week after the infection. The control is done at the limit. Another few days of exponential growth and the population of bacteria would have been uncatchable.

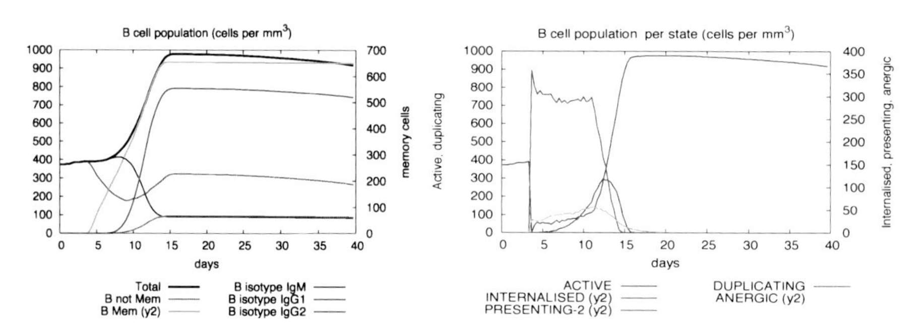

Figure 51. Left plot shows the virgin and memory B and the plot on the right shows the same population grouped by state.

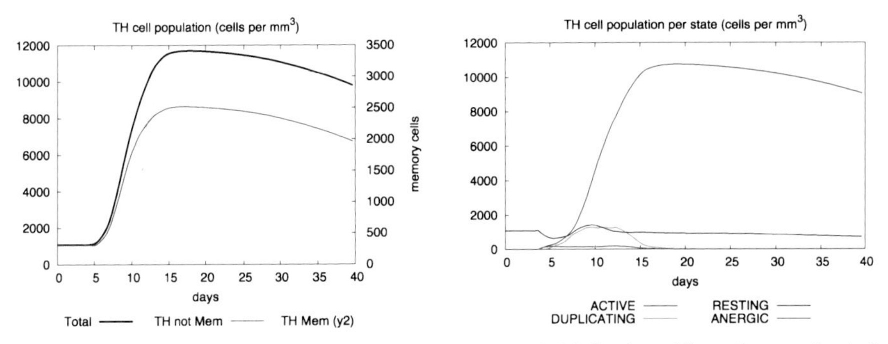

Figure 52. Helper T-cell dynamics during a bacterial infection. The plot on the left shows the virgin and memory T-cells whereas the plot on the right shows the same population by state.

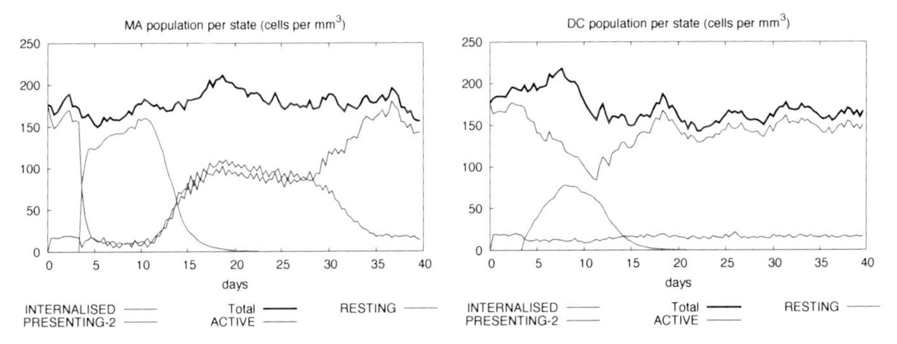

Figure 53. These plots show the dynamics of macrophages (MA, left) and dendritic cells (DC, right) during a bacterial infection. Dendritic cells are shown to do a better job of presenting the antigen. Both macrophages and dendritic cells present the captured antigen to helper T-cells to mount the immune response. Macrophages need to be activated before this work can begin.

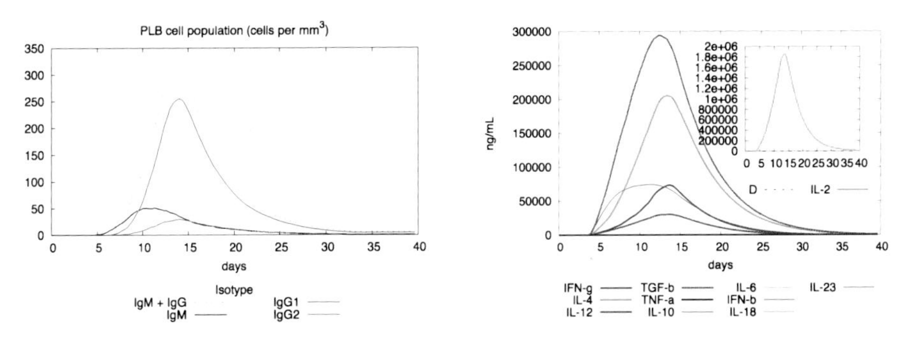

Figure 54. Plasma B cells (PLB, left) and cytokines (right) during an immune response to a fast-replicating bacterium. IgG1-producing PLB cells surpass IgM- and IgG2-producing cells. On the cytokines pattern, the presence of TGF-β and IL-4 reveals a Th2 response, but because pro-Th1 cytokines IFN-γ and IL-12 are also present, the response is not at all unbalanced.

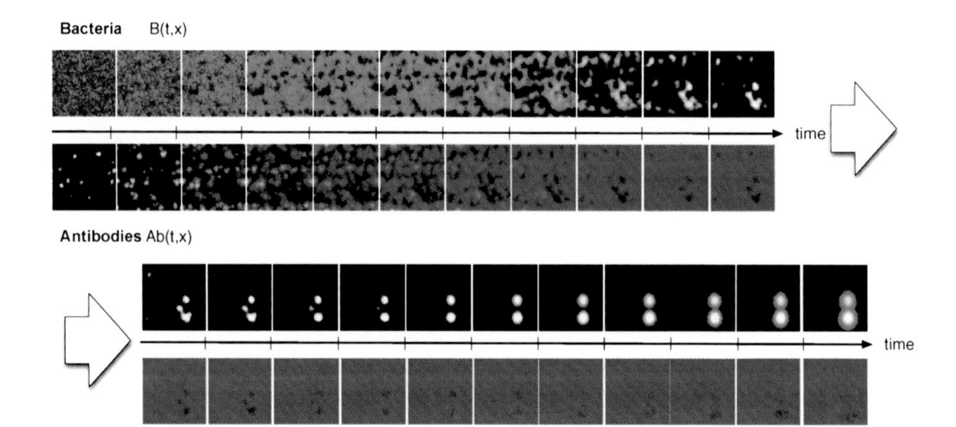

Figure 55. Spatial evolution of a bacterial infection. Twenty-two shots of the bacterial concentration, $B(t, x)$, and antibody concentration, $Ab(t, x)$, on a two-dimensional squared lattice (1000 × 1000). Dark/black areas indicate low concentrations. It is interesting to observe the formation of two spots in which bacteria survive and proliferate, notwithstanding a high concentration of antibodies.

Figure 56. Illustrations of the coronavirus (left) responsible for the SARS, the *West Nile* virus responsible for the West Nile Fever (centre) and the HIV-1 virus responsible for AIDS (right). The fine structure of the viruses is not implemented in C-ImmSim: only their most important features like replication, infectivity and mutation are represented, together with details about their life cycles.

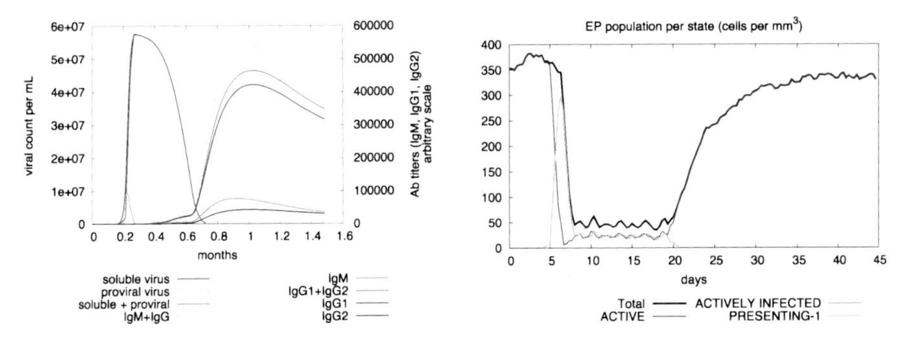

Figure 57. The virus infection triggers the production of immunoglobulins, mainly of the IgG1 type (left panel). In this plot, *soluble* means free virus while *proviral* indicates viral count inside infected cells. In the right panel, the detailed dynamics of epithelial cells (EP) as targets of the virus is shown; after a primary immune cytotoxic response (see Figure 60) which largely decreases their population, epithelial cells are kept at a low count by a continuous virus killing by blasting and by CTLs. This is visible by the large fraction (almost the totality) of cells in the state ACTIVELY INFECTED. On the other hand, a small fraction of the infected cells are able to present the virus peptides with the MHC-1 molecule on their surface.

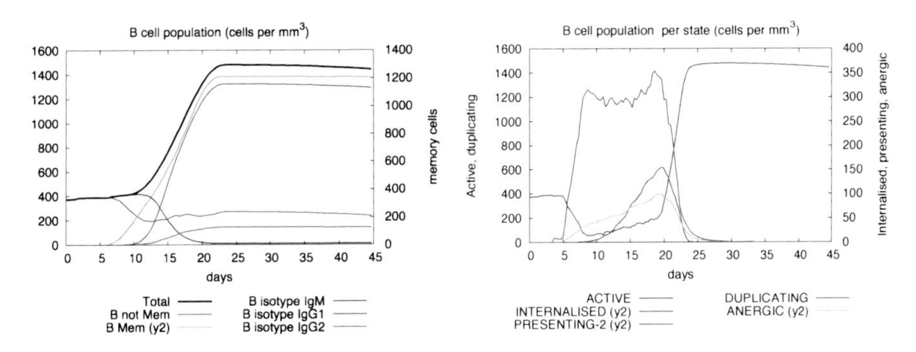

Figure 58. B lymphocytes grow in number during a viral infection. In the initial acute phase, some B-cells become anergic because of the high concentration of viral particles. Eventually, B-cells duplicate and differentiate to plasma B cells (PLB), producing antibodies.

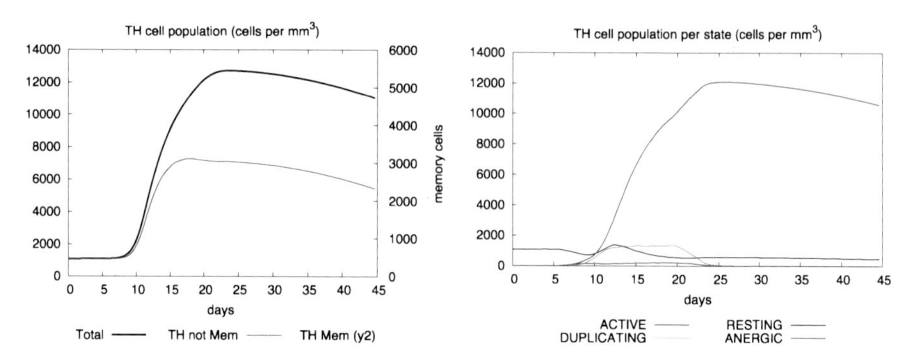

Figure 59. These are the Th cells during a viral infection. A large number of active T helper cells is deployed to eliminate the threat.

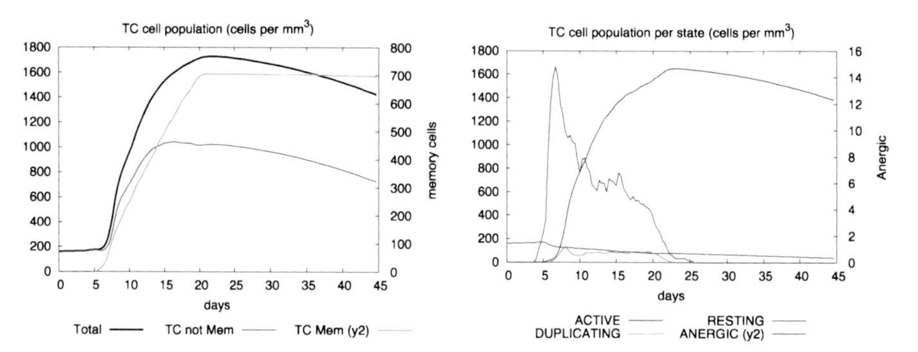

Figure 60. Cytotoxic T-cells responding to a viral infection. During the acute phase of the infection, some Tc become anergic by overstimulation. However, these anergic Tc are very few, and the majority of cytotoxic cells are actively killing infected epithelial cells.

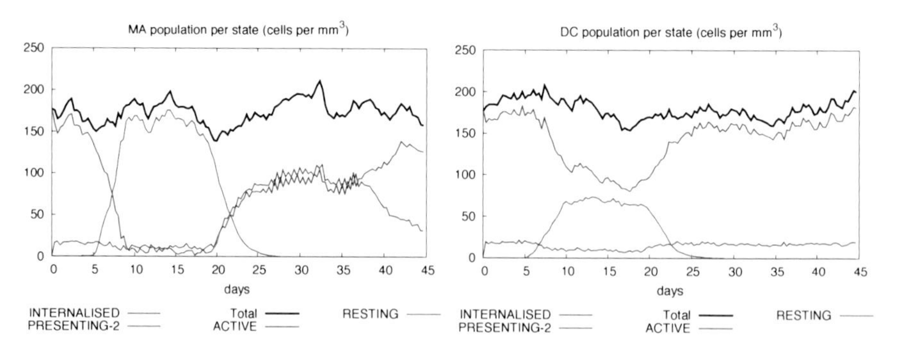

Figure 61. During the viral infection, macrophages and dendritic cells do their job by capturing soluble virus particles and presenting these antigens to helper T-cells.

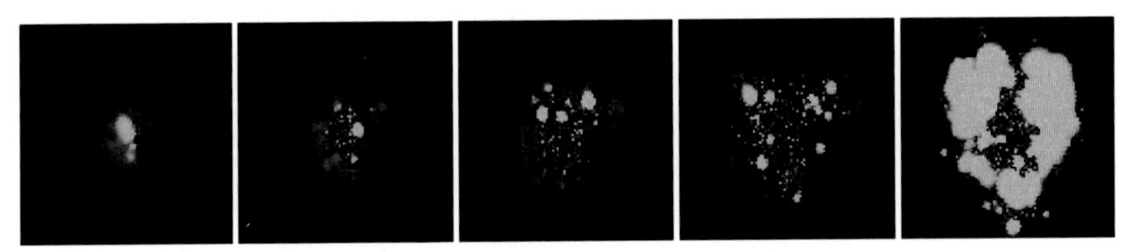

Figure 62. Five shots of a simulation of a viral infection in a lattice. Each pixel corresponds to the concentration of virus in a lattice point. Starting from a confined region, the virus spreads, mostly carried by infected cells. These burst, releasing their content and forming high-viral-density darker spots in the figure.

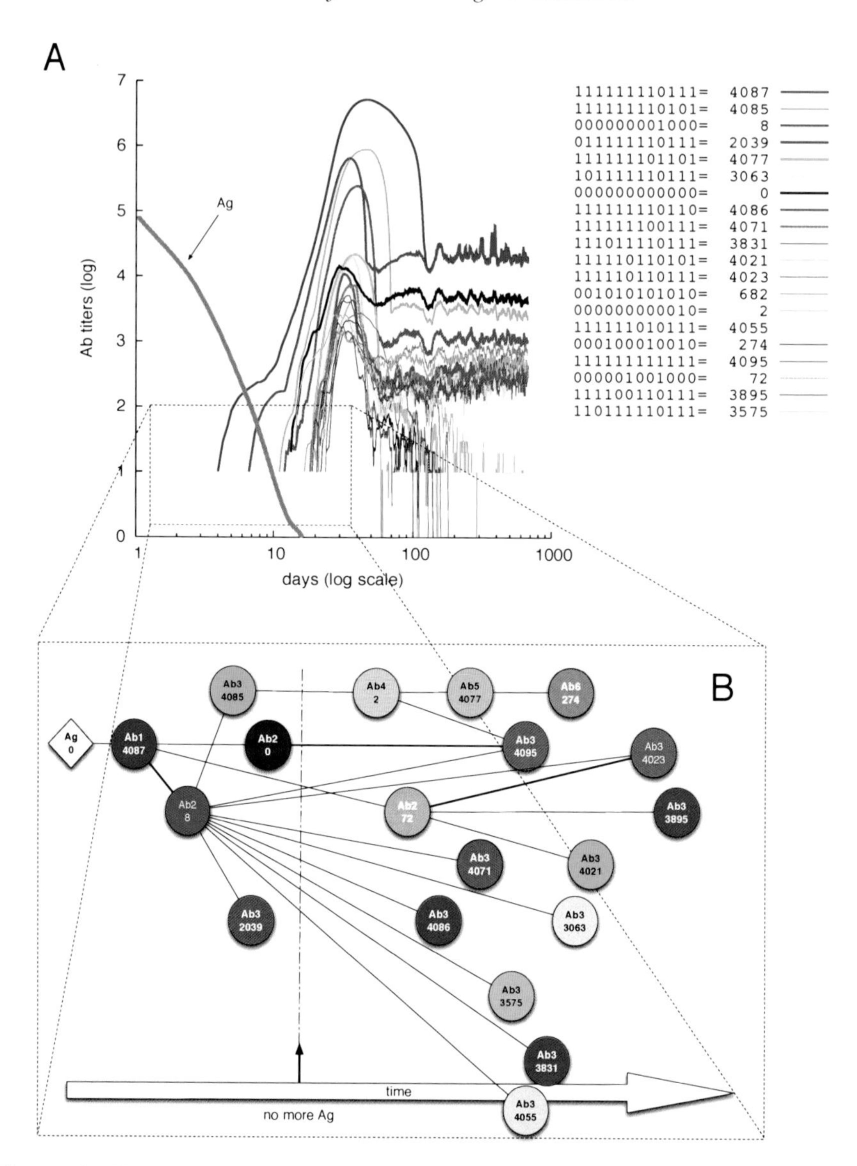

Figure 63. The idiotype network is clearly at work in this simulation result. Only the perfect match and one-bit mismatches are allowed (i.e., $m_0 = 11$). A chain of 190 idio-clones has been generated from the injected (antigen) but for simplicity, only the top twenty are shown, namely those reaching higher titers during the simulation. Some of them go to zero within about three months, whereas the remaining reach a metastable state in which they stimulate themselves, one with another. In panel B, is shown the matching relationship among the bit-strings of the idio-clones plotted in panel A. An edge between two nodes stands for a bit-string affinity above the threshold m_0. Bold edges indicate 12-bit matches and the rest, 11-bit matches. Idiotypes are shown in sequential order of appearance (left to right).

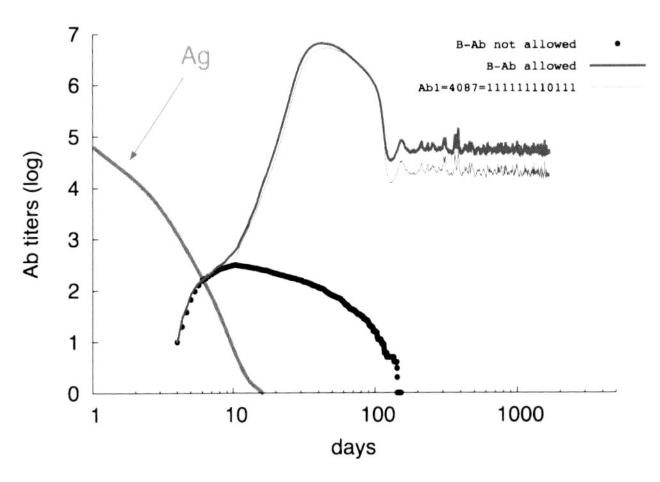

Figure 64. When B-Ab interactions are enabled, the humoral response is greater (thick orange line). Ab1 = 4087 (thin orange curve) is the dominating clone in both cases although he is just one out of 190 of the total idiotype network. The black curve shows the titer of the same antibody in case the network is not elicited. The whole humoral response consists of the Ab = 4087 only.

Chapter 5

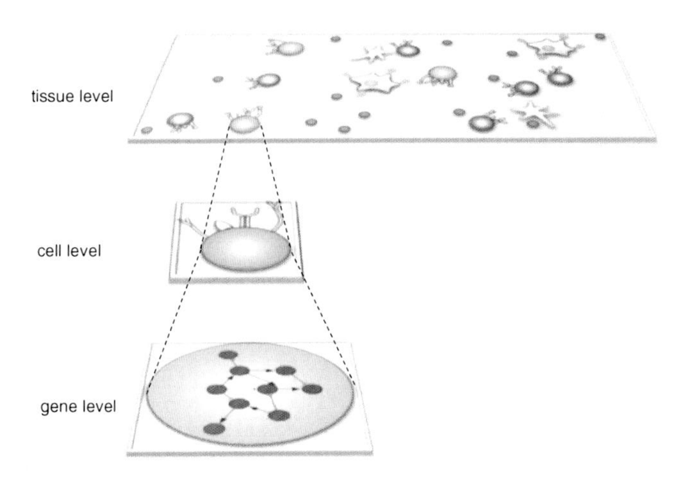

Figure 65. A multi-scale model can be constructed by simulating the gene expression dynamics by means of a Boolean gene-regulatory network coupled with the extracellular level of cytokine signaling. The cross-talk among cells represents tissue level dynamics.

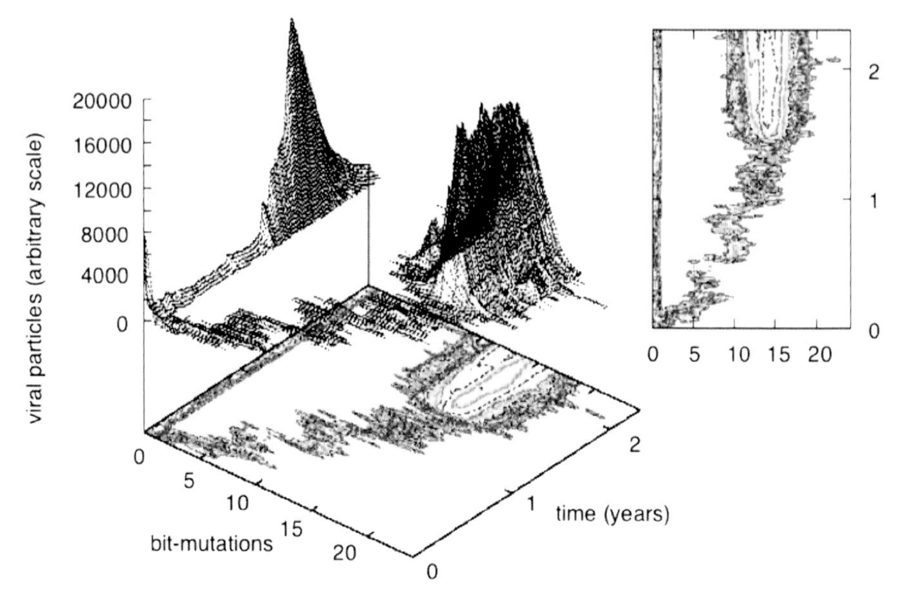

Figure 68. HIV population as a function of the number of mutations from the wild type virus. The distance is expressed in terms of bit mutations, both on the epitope and the peptide, for a total of 2 × NBIT = 24 possible mutations.

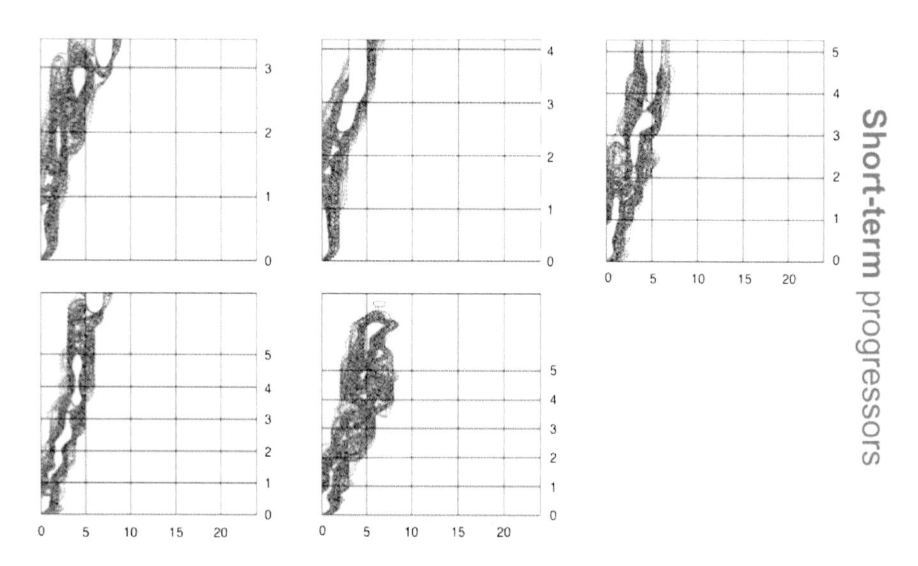

Figure 69. The evolution of the viral strains show different patterns in the short-term progressors. As in Figure 68, x-axis stands for bit-mutations and y-axis for time.

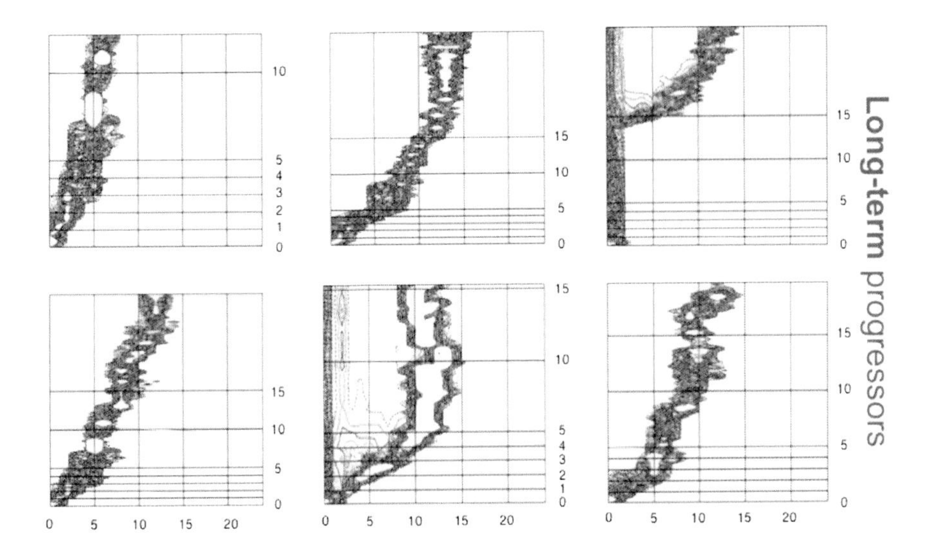

Figure 70. Long-term progressors show that the virus has a slower but more effective evolution path.

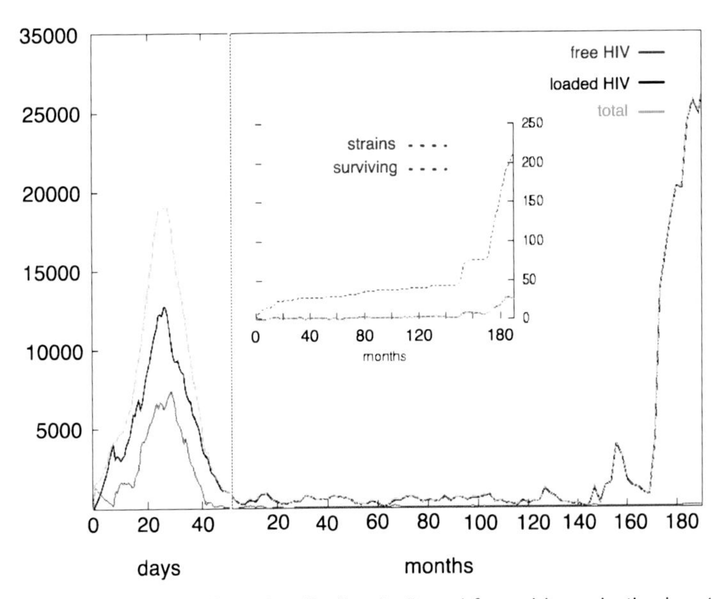

Figure 71. Virions inside infected cells (loaded) and free virions. In the inset plot are the number of generated strains and those surviving the immune selection.

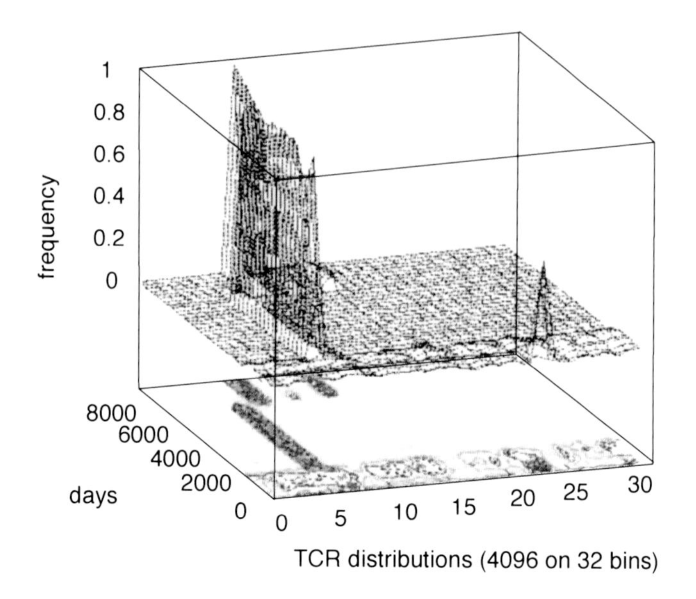

Figure 73. TCR's distribution. 4,096 bit-strings representing the potential repertoire of helper T-cell receptors have been clumped into 32 bins for clarity. Note the initial unbiased distribution quickly turning into a peak (bin~24) during the acute phase of the infection. A few memory clones (bin~7) eventually dominate the scene.

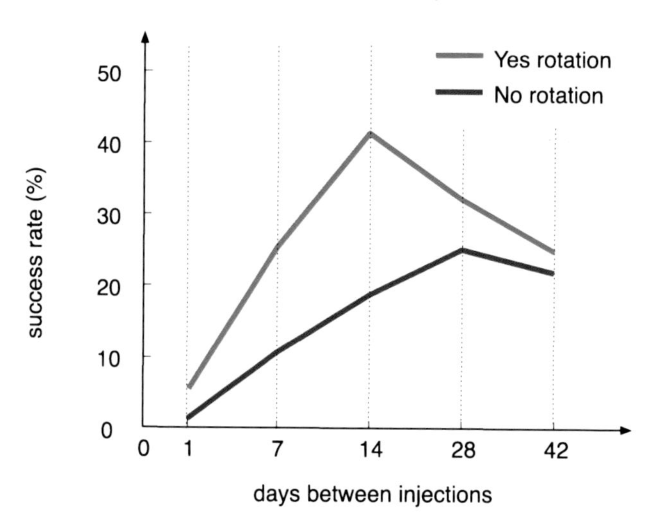

Figure 75. Percentage of success for NR and YR simulation settings for all different injection protocols tested. A successful simulation is one in which the number of tumour cells at the end of the run was lower than at the beginning. Chi-square tests indicate that for the 1, 7 and 14 days sets of simulations, the difference between NR and YR is significant (respectively 5%, 1% and 1%) whereas for the 28 and 42 days sets of simulation, it is not significant (from [136]).